Delmar's Test Preparation Series

Automobile Test

Automatic Transmission/Transaxle (Test A2)

Technical Advisor
Boyce H. Dwiggins

Delmar Publishers

an International Thomson Publishing company

Albany • Bonn • Boston • Cincinnati • Detroit • London • Madrid
Melbourne • Mexico City • New York • Pacific Grove • Paris • San Francisco
Singapore • Tokyo • Toronto • Washington

NOTICE TO THE READER

Publisher does not warrant or guarantee any of the products described herein or perform any independent analysis in connection with any of the product information contained herein. Publisher does not assume, and expressly disclaims, any obligation to obtain and include information other than that provided to it by the manufacturer.

The reader is expressly warned to consider and adopt all safety precautions that might be indicated by the activities herein and to avoid all potential hazards. By following the instructions contained herein, the reader willingly assumes all risks in connections with such instructions.

The publisher makes no representation or warranties of any kind, including but not limited to, the warranties of fitness for particular purpose or merchantability, nor are any such representations implied with respect to the material set forth herein, and the publisher takes no responsibility with respect to such material. The publisher shall not be liable for any special, consequential, or exemplary damages resulting, in whole or part, from the readers' use of, or reliance upon, this material.

Cover Design: Paul Roseneck

Delmar Staff:
Publisher: Alar Elken
Acquisitions Editor: Vernon Anthony
Editorial Assistant: Betsy Hough
Marketing Manager: Mona Caron

Online Services

Delmar Online
To access a wide variety of Delmar products and services on the World Wide Web, point your browser to:
http://www.delmar.com
or email: info@delmar.com

A service of ITP

For more information contact:

Delmar Publishers
3 Columbia Circle, Box 15015
Albany, New York 12212-5015

International Thomson Publishing Europe
Berkshire House
168-173 High Holborn
London, WC1V7AA
United Kingdom

Nelson ITP, Australia
102 Dodds Street
South Melbourne,
Victoria, 3205 Australia

Nelson Canada
1120 Birchmont Road
Scarborough, Ontario
M1K 5G4, Canada

International Thomson Publishing France
Tour Maine-Montparnasse
33 Avenue du Maine
75755 Paris Cedex 15, France

International Thomson Editores
Seneca 53
Colonia Polanco
11560 Mexico D. F. Mexico

International Thomson Publishing GmbH
Königswinterer Strasse 418
53227 Bonn
Germany

International Thomson Publishing Asia
60 Albert Street
#15-01 Albert Complex
Singapore 189969

International Thomson Publishing Japan
Hirakawa-cho Kyowa Building, 3F
2-2-1 Hirakawa-cho, Chiyoda-ku,
Tokyo 102, Japan

ITE Spain/ Paraninfo
Calle Magallanes, 25
28015-Madrid, Spain

5 6 7 8 9 10 XXX 03 02 01

ISBN 0-7668-0550-6

Contents

Section 3 Are You Sure You're Ready for A2 Test?

Section 4 An Overview of the System

Section 5 Sample Test for Practice

Section 6 Additional Test Questions for Practice

Section 7 Appendices

Preface

This book is just one of a comprehensive series designed to prepare technicians to take and pass every ASE test. Delmar's series covers all of the Automotive tests A1 through A8 as well as Advanced Engine Performance L2 and Parts Specialist P2. The series also covers the five Collision Repair tests and the eight Medium/Heavy Duty truck test.

Before any book in this series was written, Delmar staff met with and surveyed technicians and shop owners who have taken ASE tests and have used other preparatory materials. We found that they wanted, first and foremost, *lots* of practice tests and questions. Each book in our series contains a general knowledge pretest, a sample test, and additional practice questions. You will be hard-pressed to find a test prep book with more questions for you to practice with. We have worked hard to ensure that these questions match the ASE style in types of questions, quantities, and level of difficulty.

Technicians also told us that they wanted to understand the ASE test and to have practical information about what they should expect. We have provided that as well, including a history of ASE and a section devoted to helping the technician "Take and Pass Every ASE Test" with case studies, test-taking strategies, and test formats.

Finally, techs wanted refresher information and reference. Each of our books includes an overview section that is referenced to the task list. The complete task lists for each test appear in each book for the user's reference. There is also a complete glossary of terms for each booklet.

So whether you're looking for a sample test and a few extra questions to practice with or a complete introduction to ASE testing, with support for preparing thoroughly, this book series is an excellent answer.

We hope you benefit from this book and that you pass every ASE test you take!

Your comments, both positive and negative, are certainly encouraged! Please contact us at:

Automotive Editor
Delmar Publishers
3 Columbia Circle
Box 15015
Albany, NY 12212-5015

The History of ASE

History

Originally known as The National Institute for Automotive Service Excellence (NIASE), today's ASE was founded in 1972 as a non-profit, independent entity dedicated to improving the quality of automotive service and repair through the voluntary testing and certification of automotive technicians. Until that time, consumers had no way of distinguishing between competent and incompetent automotive mechanics. In the mid-1960s and early 1970s, efforts were made by several automotive industry affiliated associations to respond to this need. Though the associations were non-profit, many regarded certification test fees merely as a means of raising additional operating capital. Also, some associations, having a vested interest, produced test scores heavily weighted in the favor of its members.

NIASE

From these efforts a new independent, non-profit association, the National Institute for Automotive Service Excellence (NIASE), was established much to the credit of two educators, George R. Kinsler, Director of Program Development for the Wisconsin Board of Vocational and Adult Education in Madison, WI, and Myron H. Appel, Division Chairman at Cypress College in Cypress, CA.

Early efforts were to encourage voluntary certification in four general areas:

TEST AREA	TITLES
I. Engine	Engines, Engine Tune-Up, Block Assembly, Cooling and Lube Systems, Induction, Ignition, and Exhaust
II. Transmission	Manual Transmissions, Drive Line and Rear Axles, and Automatic Transmissions
III. Brakes and Suspension	Brakes, Steering, Suspension, and Wheels
IV. Electrical/Air Conditioning	Body/Chassis, Electrical Systems, Heating, and Air Conditioning

In early NIASE tests, Mechanic A, Mechanic B type questions were used. Over the years the trend has not changed, but in mid-1984 the term was changed to Technician A, Technician B to better emphasize sophistication of the skills needed to perform successfully in the modern motor vehicle industry. In certain tests the term used is Estimator A/B, Painter A/B, or Parts Specialist A/B. At about that same time, the logo was changed from "The Gear" to "The Blue Seal," and the organization adopted the acronym ASE for Automotive Service Excellence.

Since those early beginnings, several other related trades have been added. ASE now administers a comprehensive series of certification exams for automotive and light truck repair technicians, medium and heavy truck repair technicians, alternate fuels technicians, engine machinists, collision repair technicians, school bus repair technicians, and parts specialists.

The Series and Individual Tests

- Automotive and Light Truck Technician; consisting of: Engine Repair - Automatic Transmission/Transaxle - Manual Drive Train and Axles - Suspension and Steering - Brakes - Electrical/Electronic Systems - Heating and Air Conditioning - Engine Performance
- Medium and Heavy Truck Technician; consisting of: Gasoline Engines - Diesel Engines - Drive Train - Brakes - Suspension and Steering - Electrical/Electronic Systems - HVAC - Preventive Maintenance Inspection
- Alternate Fuels Technician; consisting of: Compressed Natural Gas Light Vehicles
- Advanced Series; consisting of: Automobile Advanced Engine Performance and Advanced Diesel Engine Electronic Diesel Engine Specialty
- Collision Repair Technician; consisting of: Painting and Refinishing - Non-Structural Analysis and Damage Repair - Structural Analysis and Damage Repair - Mechanical and Electrical Components - Damage Analysis and Estimating
- Engine Machinist Technician; consisting of: Cylinder Head Specialist - Cylinder Block Specialist - Assembly Specialist
- School Bus Repair Technician; consisting of: Body Systems and Special Equipment - Drive Train - Brakes - Suspension and Steering - Electrical/Electronic Systems - Heating and Air Conditioning
- Parts Specialist; consisting of: Automobile Parts Specialist - Medium/Heavy Truck Parts Specialist

A Brief Chronology

1970–1971 Original questions were prepared by a group of forty auto mechanics teachers from public secondary schools, technical institutes, community colleges, and private vocational schools. These questions were then professionally edited by testing specialists at Educational Testing Service (ETS) at Princeton, New Jersey, and thoroughly reviewed by training specialists associated with domestic and import automotive companies.

1971 July: About eight hundred mechanics tried out the original test questions at experimental test administrations.

1972 November and December: Initial NIASE tests administered at 163 test centers. The original automotive test series consisted of four tests containing eighty questions each. Three hours were allotted for each test. Those who passed all four tests were designated Certified General Auto Mechanic (GAM).

1973 April and May: Test 4 was increased to 120 questions. Time was extended to four hours for this test. There were now 182 test centers. Shoulder patch insignias were made available.

	November: Automotive series expanded to five tests. Heavy-Duty Truck series of six tests introduced.
1974	November: Automatic Transmission (Light Repair) test modified. Name changed to Automatic Transmission.
1975	May: Collision Repair series of two tests is introduced.
1978	May: Automotive recertification testing is introduced.
1979	May: Heavy-Duty Truck recertification testing is introduced.
1980	May: Collision Repair recertification testing is introduced.
1982	May: Test administration providers switched from Educational Testing Service (ETS) to American College Testing (ACT). Name of Automobile Engine Tune-Up test changed to Engine Performance test.
1984	May: New logo was introduced. ASE's "The Blue Seal" replaced NIASE's "The Gear." All reference to Mechanic A, Mechanic B was changed to Technician A, Technician B.
1990	November: The first of the Engine Machinist test series was introduced.
1991	May: The second test of the Engine Machinist test series was introduced. November: The third and final Engine Machinist test was introduced.
1992	May: Name of Heavy-Duty Truck Test series changed to Medium/Heavy Truck Test series.
1993	May: Automotive Parts Specialist test introduced. Collision Repair expanded to six tests. Light Vehicle Compressed Natural Gas test introduced. Limited testing begins in English-speaking provinces of Canada.
1994	May: Advanced Engine Performance Specialist test introduced.
1996	May: First three tests for School Bus Technician test series introduced. November: A Collision Repair test is added.
1997	May: A Medium/Heavy Truck test is added.
1998	May: A diesel advanced engine test is introduced: Electronic Diesel Engine Diagnosis Specialist. A test is added to the School Bus test series.

By the Numbers

Following are the approximate number of ASE technicians currently certified by category. The numbers may vary from time to time but are reasonably accurate for any given period. More accurate data may be obtained from ASE, which provides updates twice each year, in May and November after the Spring and Fall test series.

There are more than 338,000 Automotive Technicians with over 87,000 at Master Technician (MA) status. There are 47,000 Truck Technicians with over 19,000 at Master Technician (MT) status. There are 46,000 Collision Repair/Refinish Technicians with 7,300 at Master Technician (MB) status. There are 1,200 Estimators. There are 6,700 Engine Machinists with over 2,800 at Master Machinist Technician (MM) status. There are also 28,500 Automobile Advanced Engine Performance Technicians and over 2,700 School Bus Technicians for a combined total of more than 403,000 Repair Technicians. To this number, add over 22,000 Automobile Parts Specialists, and over 2,000 Truck Parts Specialists for a combined total of over 24,000 parts specialists.

There are over 6,400 ASE Technicians holding both Master Automotive Technician and Master Truck Technician status, of which 350 also hold Master Body Repair status. Almost 200 of these Master Technicians also hold Master Machinist status and five Technicians are certified in all ASE specialty areas.

Almost half of ASE certified technicians work in new vehicle dealerships (45.3%). The next greatest number work in independent garages with 19.8%. Next is tire dealerships with 9%, service stations at 6.3%, fleet shops at 5.7%, franchised volume retailers at 5.4%, paint and body shops at 4.3%, and specialty shops at 3.9%.

Of over 400,000 automotive technicians on ASE's certification rosters, almost 2,000 are female. The number of female technicians is increasing at a rate of about 20% each year. Women's increasing interest in automotive mechanics is further evidenced by the fact that, according to the National Automobile Dealers Association (NADA), they influence 80% of the decisions of the purchase of a new automobile and represent 50% of all new car purchasers. Also, it is interesting to note that 65% of all repair and maintenance service customers are female.

The typical ASE certified technician is 36.5 years of age, is computer literate, deciphers a half-million pages of technical manuals, spends one hundred hours per year in training, holds four ASE certificates, and spends about $27,000 for tools and equipment. Twenty-seven percent of today's skilled ASE certified technicians attended college, many having earned an Associate of Science degree in Automotive Technology.

ASE

ASE's mission is to improve the quality of vehicle repair and service in the United States through the testing and certification of automotive repair technicians. Prospective candidates register for and take one or more of ASE's thirty-three exams. The tests are grouped into specialties for automobile, medium/heavy truck, school bus, and collision repair technicians as well as engine machinists, alternate fuels technicians, and parts specialists.

Upon passing at least one exam and providing proof of two years of related work experience, the technician becomes ASE certified. A technician who passes a series of exams earns ASE Master Technician status. An automobile technician, for example, must pass eight exams for this recognition.

The tests, conducted twice a year at over seven hundred locations around the country, are administered by American College Testing (ACT). They stress real-world diagnostic and repair problems. Though a good knowledge of theory is helpful to the technician in answering many of the questions, there are no questions specifically on theory. Certification is valid for five years. To retain certification, the technician must be retested to renew his or her certificate.

The automotive consumer benefits because ASE certification is a valuable yardstick by which to measure the knowledge and skills of individual technicians, as well as their commitment to their chosen profession. It is also a tribute to the repair facility employing ASE certified technicians. ASE certified technicians are permitted to wear blue and white ASE shoulder insignia, referred to as the "Blue Seal of Excellence," and carry credentials listing their areas of expertise. Often employers display their technicians' credentials in the customer waiting area. Customers look for facilities that display ASE's Blue Seal of Excellence logo on outdoor signs, in the customer waiting area, in the telephone book (Yellow Pages), and in newspaper advertisements.

The tests stress repair knowledge and skill. All test takers are issued a score report. In order to earn ASE certification, a technician must pass one or more of the exams and present proof of two years of relevant hands-on work experience. ASE certifications are valid for five years, after which time technicians must retest in order to keep up with changing technology and to remain in the ASE program. A nominal registration and test fee is charged.

To become part of the team that wear ASE's Blue Seal of Excellence®, please contact:

National Institute for Automotive Service Excellence
13505 Dulles Technology Drive
Herndon, VA 20171-3421

2 Take and Pass Every ASE Test

ASE Testing

Participating in an Automotive Service Excellence (ASE) voluntary certification program gives you a chance to show your customers that you have the "know-how" needed to work on today's modern vehicles. The ASE certification tests allow you to compare your skills and knowledge to the automotive service industry's standards for each specialty area.

If you are the "average" automotive technician taking this test, you are in your mid-thirties and have not attended school for about fifteen years. That means you probably have not taken a test in many years. Some of you, on the other hand, have attended college or taken postsecondary education courses and may be more familiar with taking tests and with test-taking strategies. There is, however, a difference in the ASE test you are preparing to take and the educational tests you may be accustomed to.

Who Writes the Questions?

The questions on an educational test are generally written, administered, and graded by an educator who may have little or no practical hands-on experience in the test area. The questions on all ASE tests are written by service industry experts familiar with all aspects of the subject area. ASE questions are entirely job-related and designed to test the skills that you need to know on the job.

The questions originate in an ASE "item-writing" workshop where service representatives from domestic and import automobile manufacturers, parts and equipment manufacturers, and vocational educators meet in a workshop setting to share their ideas and translate them into test questions. Each test question written by these experts is reviewed by all of the members of the group. The questions deal with the practical problems of diagnosis and repair that are experienced by technicians in their day-to-day hands-on work experiences.

All of the questions are pretested and quality-checked in a nonscoring section of tests by a national sample of certifying technicians. The questions that meet ASE's high standards of accuracy and quality are then included in the scoring sections of future tests. Those questions that do not pass ASE's stringent test are sent back to the workshop or are discarded. ASE's tests are monitored by an independent proctor and are administered and machine-scored by an independent provider, American College Testing (ACT). All ASE tests have a three-year revision cycle.

Testing

If you think about it, we are actually tested on about everything we do. As infants, we were tested to see when we could turn over and crawl, later when we could walk or talk. As adolescents, we were tested to determine how well we learned the material presented in school and in how we demonstrated our accomplishments on the athletic field. As working adults, we are tested by our supervisors on how well we have completed an assignment or project. As nonworking adults, we are tested by our family on everyday activities, such as housekeeping or preparing a meal. Testing, then, is one of those facts of life that begins in the cradle and follows us to the grave.

Testing is an important fact of life that helps us to determine how well we have learned our trade. Also, tests often help us to determine what opportunities will be available to us in the future. To become ASE certified, we are required to take a test in every subject in which we wish to be recognized.

Be Test-Wise

In spite of the widespread use of tests, most technicians are not very test-wise. An ability to take tests and score well is a skill that must be acquired. Without this knowledge, the most intelligent and prepared technician may not do well on a test.

We will discuss some of the basic procedures necessary to follow in order to become a test-wise technician. Assume, if you will, that you have done the necessary study and preparation to score well on the ASE test.

Different approaches should be used for taking different types of tests. The different basic types of tests include: essay, objective, multiple choice, fill in the blank, true-false, problem solving, and open book. All ASE tests are of the four-part multiple-choice type.

Before discussing the multiple-choice type test questions, however, there are a few basic principles that should be followed before taking any test.

Before the Test

Do not arrive late. Always arrive well before your test is scheduled to begin. Allow ample time for the unexpected, such as traffic problems, so you will arrive on time and avoid the unnecessary anxiety of being late.

Always be certain to have plenty of supplies with you. For an ASE test, three or four sharpened soft lead (#2) pencils, a pocket pencil sharpener, erasers, and a watch are all that are required.

Do not listen to pretest chatter. When you arrive early, you may hear other technicians testing each other on various topics or making their best guess as to the probable test questions. At this time, it is too late to add to your knowledge. Also the rhetoric may only confuse you. If you find it bothersome, take a walk outside the test room to relax and loosen up.

Read and listen to all instructions. It is important to read and listen to the instructions. Make certain that you know what is expected of you. Listen carefully to verbal instructions and pay particular attention to any written instructions on the test paper. Do not dive into answering questions only to find out that you have answered the wrong question by not following instructions carefully. It is difficult to make a high score on a test if you answer the wrong questions.

These basic principles have been violated in most every test ever given. Try to remember them. They are essential for success.

Objective Tests

A test is called an objective test if the same standards and conditions apply to everyone taking the test and there is only one correct answer to each question. Objective tests primarily measure your ability to recall information. A well-designed objective test can also test your ability to understand, analyze, interpret, and apply your knowledge. Objective tests include true-false, multiple choice, fill in the blank, and matching questions.

Objective questions, not generally encountered in a classroom setting, are frequently used in standardized examinations. Objective tests are easy to grade and also reduce the amount of paperwork necessary to administer. The objective tests are used in entry-level programs or when very large numbers are being tested. ASE's tests consist exclusively of four-part multiple-choice objective questions in all of their tests.

Taking an Objective Test

The principles of taking an objective test are somewhat different from those used in other types of tests. You should first quickly look over the test to determine the number of questions, but do not try to read through all of the questions. In an ASE test, there are usually between forty and eighty questions, depending on the subject matter. Read through each question before marking your answer. Answer the questions in the order they appear on the test. Leave the questions blank that you are not sure of and move on to the next question. You can return to those unanswered questions after you have finished the others. They may be easier to answer at a later time after your mind has had additional time to consider them on a subconscious level. In addition, you might find information in other questions that will help you to answer some of them.

Do not be obsessed by the apparent pattern of responses. For example, do not be influenced by a pattern like **d**, **c**, **b**, **a**, **d**, **c**, **b**, **a** on an ASE test.

There is also a lot of folk wisdom about taking objective tests. For example, there are those who would advise you to avoid response options that use certain words such as *all, none, always, never, must,* and *only,* to name a few. This, they claim, is because nothing in life is exclusive. They would advise you to choose response options that use words that allow for some exception, such as *sometimes, frequently, rarely, often, usually, seldom,* and *normally.* They would also advise you to avoid the first and last option (A and D) because test writers, they feel, are more comfortable if they put the correct answer in the middle (B and C) of the choices. Another recommendation often offered is to select the option that is either shorter or longer than the other three choices because it is more likely to be correct. Some would advise you to never change an answer since your first intuition is usually correct.

Although there may be a grain of truth in this folk wisdom, ASE test writers try to avoid them and so should you. There are just as many **A** answers as there are **B** answers, just as many **D** answers as **C** answers. As a matter of fact, ASE tries to balance the answers at about 25 percent per choice **A**, **B**, **C**, and **D**. There is no intention to use "tricky" words, such as outlined above. Put no credence in the opposing words "sometimes" and "never," for example. When used in an ASE type question, one or both may be correct; one or both may be incorrect.

There are some special principles to observe on multiple-choice tests. These tests are sometimes challenging because there are often several choices that may seem possible, and it may be difficult to decide on the correct choice. The best strategy, in this case, is to first determine the correct answer before looking at the options. If you see the answer you decided on, you should still examine the options to make sure that none seem more correct than yours. If you do not know or are not sure of the answer, read each option very carefully and try to eliminate those options that you know to be wrong. That way, you can often arrive at the correct choice through a process of elimination.

If you have gone through all of the test and you still do not know the answer to some of the questions, then guess. Yes, guess. You then have at least a 25 percent chance of being correct. If you leave the question blank, you have no chance. In ASE tests, there is no penalty for being wrong. As the late President Franklin D. Roosevelt once advised a group of students, "It is common sense to take a method and try it. If it fails, admit it frankly and try another. But above all, try something."

During the Test

Mark your bubble sheet clearly and accurately. One of the biggest problems an adult faces in test-taking, it seems, is in placing an answer in the correct spot on a bubble sheet. Make certain that you mark your answer for, say, question 21, in the space on the bubble sheet designated for the answer for question 21. A correct response in the wrong bubble will probably be wrong. Remember, the answer sheet is machine scored and can only "read" what you have bubbled in. Also, do not bubble in two answers for the same question. For example, if you feel the answer to a particular question is **A** but think it may be **C**, do not bubble in both choices. Even if either **A** or **C** is correct, a double answer will score as an incorrect answer. It's better to take a chance with your best guess.

Review Your Answers

If you finish answering all of the questions on a test ahead of time, go back and review the answers of those questions that you were not sure of. You can often catch careless errors by using the remaining time to review your answers.

Don't Be Distracted

At practically every test, some technicians will invariably finish ahead of time and turn their papers in long before the final call. Do not let them distract or intimidate you. Either they knew too little and could not finish the test, or they were very self-confident and thought they knew it all. Perhaps they were trying to impress the proctor or other technicians about how much they know. Often you may hear them later talking about the information they knew all the while but forgot to respond on their answer sheet.

Use Your Time Wisely

It is not wise to use less than the total amount of time that you are allotted for a test. If there are any doubts, take the time for review. Any product can usually be made better with some additional effort. A test is no exception. It is not necessary to turn in your test paper until you are told to do so.

Don't Cheat

Some technicians may try to use a "crib sheet" during a test. Others may attempt to read answers from another technician's paper. If you do that, you are unquestionably assuming that someone else has a correct answer. You probably know as much, maybe more, than anyone else in the test room. Trust yourself. If you're still not convinced, think of the consequences of being caught. Cheating is foolish. If you are caught, you have failed the test.

Be Confident

The first and foremost principle in taking a test is that you need to know what you are doing, to be test-wise. It will now be presumed that you are a test-wise technician and are now ready for some of the more obscure aspects of test-taking.

An ASE-style test requires that you use the information and knowledge at your command to solve a problem. This generally requires a combination of information similar to the way you approach problems in the real world. Most problems, it seems, typically do not fall into neat textbook cases. New problems are often difficult to handle, whether they are encountered inside or outside the test room.

An ASE test also requires that you apply methods taught in class as well as those learned on the job to solve problems. These methods are akin to a well-equipped tool box in the hands of a skilled technician. You have to know what tools to use in a particular situation, and you must also know how to use them. In an ASE test, you will need to be able to demonstrate that you are familiar with and know how to use the tools.

You should begin a test with a completely open mind. At times, however, you may have to move out of your normal way of thinking and be creative to arrive at a correct answer. If you have diligently studied for at least one week prior to the test, you have bombarded your mind with a wide assortment of information. Your mind will be working with this information on a subconscious level, exploring the interrelationships among various facts, principles, and ideas. This prior preparation should put you in a creative mood for the test.

In order to reach your full potential, you should begin a test with the proper mental attitude and a high degree of self-confidence. You should think of a test as an opportunity to document how much you know about the various tasks in your chosen profession. If you have been diligently studying the subject matter, you will be able to take your test in serenity because your mind will be well organized. If you are confident, you are more likely to do well because you have the proper mental attitude. If, on the other hand, your confidence is low, you are bound to do poorly. It is a self-fulfilling prophecy.

Perhaps you have heard athletic coaches talk about the importance of confidence when competing in sports. Mental confidence helps an athlete to perform at the highest level and gain an advantage over competitors. Taking a test is much like an athletic event. You are competing against yourself, in a certain sense, because you will be trying to approach perfection in determining your answers. As in any competition, you should aim your sights high and be confident that you can reach the apex.

Anxiety and Fear

Many technicians experience anxiety and fear at the very thought of taking a test. Many worry, become nervous, and even become ill at test time because of the fear of failure. Many often worry about the criticism and ridicule that may come from their employer, relatives, and peers. Some worry about taking a test because they feel that the stakes are very high. Those who spent a great amount of time studying may feel they must get a high grade to justify their efforts. The thought of not doing well can result in unnecessary worry. They become so worried, in fact, that their reasoning and thinking ability is impaired, actually bringing about the problem they wanted to avoid.

The fear of failure should not be confused with the desire for success. It is natural to become "psyched-up" for a test in contemplation of what is to come. A little emotion can provide a healthy flow of adrenaline to peak your senses and hone your mental ability. This improves your performance on the test and is a very different reaction from fear.

Most technician's fears and insecurities experienced before a test are due to a lack of self-confidence. Those who have not scored well on previous tests or have no confidence in their preparation are those most likely to fail. Be confident that you will do well on your test and your fears should vanish. You will know that you have done everything possible to realize your potential.

Getting Rid of Fear

If you have previously experienced fear of taking a test, it may be difficult to change your attitude immediately. It may be easier to cope with fear if you have a better understanding of what the test is about. A test is merely an assessment of how much the technician knows about a particular task area. Tests, then, are much less threatening when thought of in this manner. This does not mean, however, that you should lower your self-esteem simply because you performed poorly on a test.

You can consider the test essentially as a learning device, providing you with valuable information to evaluate your performance and knowledge. Recognize that no one is perfect. All humans make mistakes. The idea, then, is to make mistakes before the test, learn from them, and avoid repeating them on the test. Fortunately, this is not as difficult as it seems. Practical questions in this study guide include the correct answers to consider if you have made mistakes on the practice test. You should learn where you went wrong so you will not repeat them in the ASE test. If you learn from your mistakes, the stage is set for future growth.

If you understood everything presented up until now, you have the knowledge to become a test-wise technician, but more is required. To be a test-wise technician, you not only have to practice these principles, you have to diligently study in your task area.

Effective Study

The fundamental and vital requirement to induce effective study is a genuine and intense desire to achieve. This is more basic than any rule or technique that will be given here. The key requirement, then, is a driving motivation to learn and to achieve.

If you wish to study effectively, first develop a desire to master your studies and sincerely believe that you will master them. Everything else is secondary to such a desire.

First, build up definite ambitions and ideals toward which your studies can lead. Picture the satisfaction of success. The attitude of the technician may be transformed from merely getting by to an earnest and energetic effort. The best direct stimulus to change may involve nothing more than the deliberate planning of your time. Plan time to study.

Another drive that creates positive study is an interest in the subject studied. As an automotive technician, you can develop an interest in studying particular subjects if you follow these four rules:

1. Acquire information from a variety of sources. The greater your interest in a subject, the easier it is to learn about it. Visit your local library and seek books on the subject you are studying. When you find something new or of interest, make inexpensive photocopies for future study.
2. Merge new information with your previous knowledge. Discover the relationship of new facts to old known facts. Modern developments in automotive technology take on new interest when they are seen in relation to present knowledge.
3. Make new information personal. Relate the new information to matters that are of concern to you. The information you are now reading, for example, has interest to you as you think about how it can help.
4. Use your new knowledge. Raise questions about the points made by the book. Try to anticipate what the next steps and conclusions will be. Discuss this new knowledge, particularly the difficult and questionable points, with your peers.

You will find that when you study with eager interest, you will discover it is no longer work. It is pleasure and you will be fascinated in what you study. Studying can be like reading a novel or seeing a movie that overcomes distractions and requires no effort or willpower. You will discover that the positive relationship between interest and effort works both ways. Even though you perhaps began your studies with little or no interest, simply staying with it helped you to develop an interest in your studies.

Obviously, certain subject matter studies are bound to be of little or no interest, particularly in the beginning. Parts of certain studies may continue to be uninteresting. An honest effort to master those subjects, however, nearly always brings about some level of interest. If you appreciate the necessity and reward of effective studying, you will rarely be disappointed. Here are a few important hints for gaining the determination that is essential to carrying good conclusions into actual practice.

Make Study Definite

Decide what is to be studied and when it is to be studied. If the unit is discouragingly long, break it into two or more parts. Determine exactly what is involved in the first part and learn that. Only then should you proceed to the next part. Stick to a schedule.

The Urge to Learn

Make clear to yourself the relation of your present knowledge to your study materials. Determine the relevance with regard to your long-range goals and ambitions.

Turn your attention away from real or imagined difficulties as well as other things that you would rather be doing. Some major distractions are thoughts of other duties and of disturbing problems. These distractions can usually be put aside, simply shunted off by listing them in a notebook. Most technicians have found that by writing interfering thoughts down, their minds are freed from annoying tensions.

Adopt the most reasonable solution you can find or seek objective help from someone else for personal problems. Personal problems and worry are often causes of ineffective study. Sometimes there are no satisfactory solutions. Some manage to avoid the problems or to meet them without great worry. For those who may wish to find better ways of meeting their personal problems, the following suggestions are offered:

1. Determine as objectively and as definitely as possible where the problem lies. What changes are needed to remove the problem, and which changes, if any, can be made? Sometimes it is wiser to alter your goals than external conditions. If there is no perfect solution, explore the others. Some solutions may be better than others.
2. Seek an understanding confidant who may be able to help analyze and meet your problems. Very often, talking over your problems with someone in whom you have confidence and trust will help you to arrive at a solution.
3. Do not betray yourself by trying to evade the problem or by pretending that it has been solved. If social problem distractions prevent you from studying or doing satisfactory work, it is better to admit this to yourself. You can then decide what can be done about it.

Once you are free of interferences and irritations, it is much easier to stay focused on your studies.

Concentrate

To study effectively, you must concentrate. Your ability to concentrate is governed, to a great extent, by your surroundings as well as your physical condition. When absorbed in study, you must be oblivious to everything else around you. As you learn to concentrate and study, you must also learn to overcome all distractions. There are three kinds of distractions you may face:

1. Distractions in the surrounding area, such as motion, noise, and the glare of lights. The sun shining through a window on your study area, for example, can be very distracting.

   Some technicians find that, for effective study, it is necessary to eliminate visual distractions as well as noises. Others find that they are able to tolerate moderate levels of auditory or visual distraction.

   Make sure your study area is properly lighted and ventilated. The lighting should be adequate but should not shine directly into your eyes or be visible out of the corner of your eye. Also, try to avoid a reflection of the lighting on the pages of your book.

   Whether heated or cooled, the environment should be at a comfortable level. For most, this means a temperature of 78°F–80°F (25.6°C–26.7°C) with a relative humidity of 45 to 50 percent.

2. Distractions arising from your body, such as a headache, fatigue, and hunger. Be in good physical condition. Eat wholesome meals at regular times. Try to eat with your family or friends whenever possible. Meal time should be your recreational period. Do not eat a heavy meal for lunch, and do not resume studies immediately after eating lunch. Just after lunch, try to get some regular exercise, relaxation, and recreation. A little exercise on a regular basis is much more valuable than a lot of exercise only on occasion.
3. Distractions of irrelevant ideas, such as how to repair the garden gate, when you are studying for an automotive-related test.

The problems associated with study are no small matter. These problems of distractions are generally best dealt with by a process of elimination. A few important rules for eliminating distractions follow.

Get Sufficient Sleep

You must get plenty of rest even if it means dropping certain outside activities. Avoid cutting in on your sleep time; you will be rewarded in the long run. If you experience difficulty going to sleep, do something to take your mind off your work and try to relax before going to bed. Some suggestions that may help include a little reading, a warm bath, a short walk, a conversation with a friend, or writing that overdue letter to a distant relative. If sleeplessness is an ongoing problem, consult a physician. Do not try any of the sleep remedies on the market, particularly if you are on medication, without approval of your physician.

If you still have difficulty studying, a final rule may help. Sit down in a favorable place for studying, open your books, and take out your pencil and paper. In a word, go through the motions.

Arrange Your Area

Arrange your chair and work area. To avoid strain and fatigue, whenever possible, shift your position occasionally. Try to be comfortable; however, avoid being too comfortable. It is nearly impossible to study rigorously when settled back in a large easy chair or reclining leisurely on a sofa.

When studying, it is essential to have a plan of action, a time to work, a time to study, and a time for pleasure. If you schedule your day and adhere to the schedule, you will eliminate most of your efforts and worries. A plan that is followed, then, soon becomes the easy and natural routine of the day. Most technicians find it useful to have a definite place and time to study. A particular table and chair should always be used for study and intellectual work. This place will then come to mean study. To be seated in that particular location at a regular scheduled time will automatically lead you to assume a readiness for study.

Don't Daydream

Daydreaming or mind-wandering is an enemy of effective study. Daydreaming is frequently due to an inadequate understanding of words. Use the Glossary or a dictionary to look up the troublesome word. Another frequent cause of daydreaming is a deficient background in the present subject matter. When this is the problem, go back and review the subject matter to obtain the necessary foundation. Just one hour of concentrated study is equivalent to ten hours with frequent lapses of daydreaming. Be on guard against mind-wandering, and pull yourself back into focus on every occasion.

Study Regularly

A system of regularity in study is believed by many scholars to be the secret of success. The daily time schedule must, however, be determined on an individual basis. You must decide how many hours of each day you can devote to your studies. Few technicians really are aware of where their leisure time is spent. An accurate account of how your days are presently being spent is an important first step toward creating an effective daily schedule.

WEEKLY SCHEDULE							
	SUN	MON	TUES	WED	THU	FRI	SAT
6:00							
6:30							
7:00							
7:30							
8:00							
8:30							
9:00							
9:30							
10:00							
10:30							
11:00							
11:30							
NOON							
12:30							
1:00							
1:30							
2:00							
2:30							
3:00							
3:30							
4:00							
4:30							
5:00							
5:30							
6:00							
6:30							
7:00							
7:30							
8:00							
8:30							
9:00							
9:30							
10:00							
10:30							
11:00							
11:30							

The convenient form is for keeping an hourly record of your week's activities. If you fill in the schedule each evening before bedtime, you will soon gain some interesting and useful facts about yourself and your use of your time. If you think over the causes of wasted time, you can determine how you might better spend your time. A practical schedule can be set up by using the following steps:

Mark your fixed commitments, such as work, on your schedule. Be sure to include classes and clubs. Do you have sufficient time left? You can arrive at an estimate of the time you need for studying by counting the hours used during the present week. An often used formula, if you are taking classes, is to multiply the number of hours you spend in class by two. This provides time for class studies. This is then added to your work hours. Do not forget time allocation for travel.

Fill in your schedule for meals and studying. Use as much time as you have available during the normal workday hours. Do not plan, for example, to do all of your studying between 11:00 pm and 1:00 am. Try to select a time for study that you can use every day without interruption. You may have to use two or perhaps three different study periods during the day.

List the things you need to do within a time period. A one-week time frame seems to work well for most technicians. The question you may ask yourself is: "What do I need to do to be able to walk into the test next week, or next month, prepared to pass?"

Break down each task into smaller tasks. The amount of time given to each area must also be settled. In what order will you tackle your schedule? It is best to plan the approximate time for your assignments and the order in which you will do them. In this way, you can avoid the difficulties of not knowing what to do first and of worrying about the other things you should be doing.

List your tasks in the empty spaces on your schedule. Keep some free time unscheduled so you can deal with any unexpected events, such as a dental appointment. You will then have a tentative schedule for the following week. It should be flexible enough to allow some units to be rearranged if necessary. Your schedule should allow time off from your studies. Some use the promise of a planned recreational period as a reward for motivating faithfulness to a schedule. You will more likely lose control of your schedule if it is packed too tightly.

Keep a Record

Keep a record of what you actually do. Use the knowledge you gain by keeping a record of what you are actually doing so you can create or modify a schedule for the following week. Be sure to give yourself credit for movement toward your goals and objectives. If you find that you can not study productively at a particular hour, modify your schedule so as to correct that problem.

Scoring the ASE Test

You can gain a better perspective about tests if you know and understand how they are scored. ASE's tests are scored by American College Testing (ACT), a non-partial, non-biased organization having no vested interest in ASE or in the automotive industry. Each question carries the same weight as any other question. For example, if there are fifty questions, each is worth 2 percent of the total score. The passing grade is 70 percent. That means you must correctly answer thirty-five of the fifty questions to pass the test.

Understand the Test Results

The test results can tell you:

- where your knowledge equals or exceeds that needed for competent performance, or
- where you might need more preparation.

The test results *cannot* tell you:

- how you compare with other technicians, or
- how many questions you answered correctly.

Your ASE test score report will show the number of correct answers you got in each of the content areas. These numbers provide information about your performance in each area of the test. However, because there may be a different number of questions in each area of the test, a high percentage of correct answers in an area with few questions may not offset a low percentage in an area with many questions.

It may be noted that one does not "fail" an ASE test. The technician that does not pass is simply told "More Preparation Needed." Though large differences in percentages may indicate problem areas, it is important to consider how many questions were asked in each area. Since each test evaluates all phases of the work involved in a service specialty, you should be prepared in each area. A low score in one area could keep you from passing an entire test.

Note that a typical test will contain the number of questions indicated above each content area's description. For example:

Automatic Transmission/Transaxle (Test A2)

Content Area	Questions	Percent of Test
A. General Transmission/Transaxle Diagnosis	24	48%
1. Mechanical/Hydraulic Systems (12)		
2. Electronic Systems (12)		
B. Transmission/Transaxle Maintenance and Adjustment	5	10%
C. In-Vehicle Transmission/Transaxle Repair	10	20%
D. Off-Vehicle Transmission/Transaxle Repair	11	22%
1. Removal, Disassembly, and Assembly (3)		
2. Gear Train, Shafts, Bushings, Oil Pump, and Case (4)		
3. Friction and Reaction Units (4)		
Total	*50	100%

***Note:** The test could contain up to ten additional questions that are included for statistical research purposes only. Your answers to these questions will not affect your score, but since you do not know which ones they are, you should answer all questions in the test. The five-year Recertification Test will cover the same content areas as those listed above. However, the number of questions in each content area of the Recertification Test will be reduced by about one-half.*

"Average"

There is no such thing as average. You cannot determine your overall test score by adding the percentages given for each task area and dividing by the number of areas. It doesn't work that way because there generally are not the same number of questions in each task area. A task area with twenty questions, for example, counts more toward your total score than a task area with ten questions.

So, How Did You Do?

Your test report should give you a good picture of your results and a better understanding of your task areas of strength and weakness.

If you fail to pass the test, you may take it again at any time it is scheduled to be administered. You are the only one who will receive your test score. Test scores will not be given over the telephone by ASE nor will they be released to anyone without your written permission.

3 Are You Sure You're Ready for A2 Test?

Pretest

The purpose of this pretest is to determine the amount of review that you may require prior to taking the ASE Automobile test: Automatic Transmission/Transaxle (Test A2). If you answer all of the pretest questions correctly, complete the sample test in section 5 along with the additional test questions in section 6.

If two or more of your answers to the pretest questions are wrong, study section 4: An Overview of the System before continuing with the sample test and additional test questions.

The pretest answers and selected explanations are located at the end of the pretest.

1. Technician A says low engine vacuum will cause a vacuum modulator to sense a load condition when it actually is not present, causing delayed and harsh shifts. Technician B says poor engine performance can cause delayed shifts through the action of the TV assembly. Who is right?
   A. A only
   B. B only
   C. Both A and B
   D. Neither A nor B
2. Technician A says abnormal noises from a transmission will never be caused by faulty clutches or bands. Technician B says transmission noises are typically caused by a faulty torque converter. Who is right?
   A. A only
   B. B only
   C. Both A and B
   D. Neither A nor B
3. Technician A says transmissions originally equipped with Teflon™ seals must be refitted with Teflon™ seals during an overhaul. Technician B says a press is needed for the installation of Teflon™ seals. Who is right?
   A. A only
   B. B only
   C. Both A and B
   D. Neither A nor B
4. Technician A says the oil pumps in most transmissions are press fit into the case. Technician B says the oil pumps in most transmissions are bolted to the transmission case. Who is right?
   A. A only
   B. B only
   C. Both A and B
   D. Neither A nor B

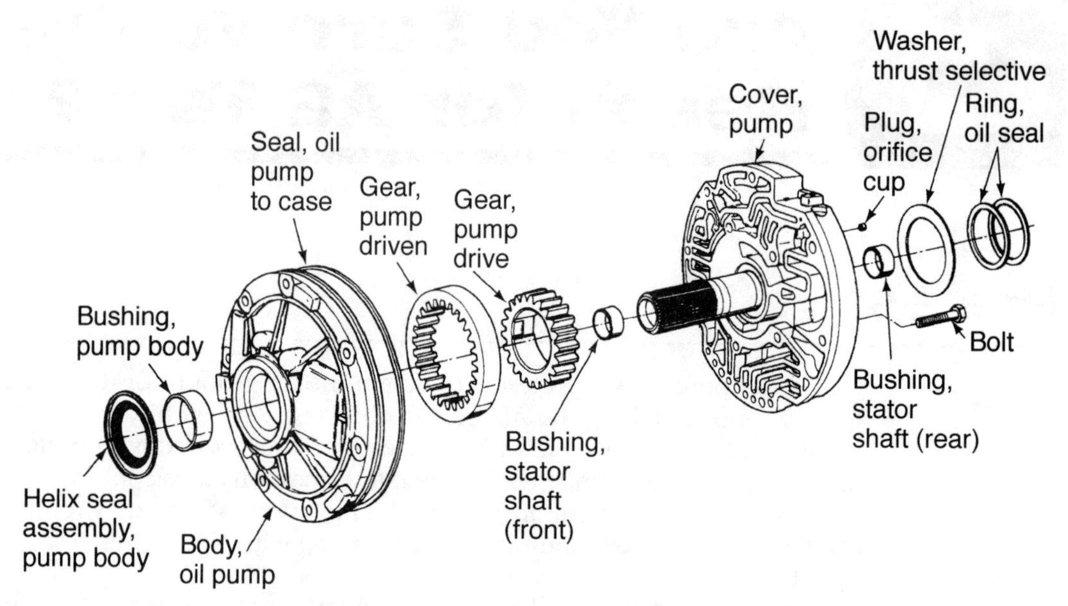

5. While assembling a transaxle, shown above, Technician A reuses all seals unless they are damaged. Technician B lubricates all seals and bearings with clean bearing grease before installing them. Who is right?
   A. A only
   B. B only
   C. Both A and B
   D. Neither A nor B
6. While checking the stall speed of a torque converter, Technician A maintains full throttle power until the engine stalls. Technician B manually engages the torque converter clutch (TCC) before running the test. Who is right?
   A. A only
   B. B only
   C. Both A and B
   D. Neither A nor B
7. While discussing universal joints, Technician A says they minimize vibrations caused by the power pulses of the engine. Technician B says that they allow the driveshaft to change angles in response to movements of the vehicle's suspension and rear axle assembly. Who is right?
   A. A only
   B. B only
   C. Both A and B
   D. Neither A nor B
8. While discussing front wheel drive (FWD) vehicles, Technician A says the differential is normally part of the transaxle assembly. Technician B says the drive axles extend from the sides of the transaxle to the drive wheels. Who is right?
   A. A only
   B. B only
   C. Both A and B
   D. Neither A nor B

9. Technician A says the rear hub of the torque converter is bolted to the flexplate. Technician B says the flexplate is designed to be flexible enough to allow the front of the converter to move forward or backward if it expands or contracts because of heat or pressure. Who is right?
   A. A only
   B. B only
   C. Both A and B
   D. Neither A nor B

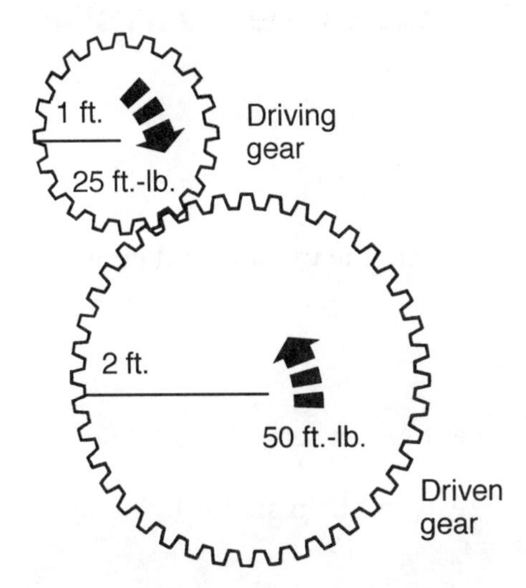

10. In the figure above, Technician A says that when a small gear drives a larger gear, output torque is increased and output speed is decreased. Technician B says that when a large gear drives a smaller gear, output torque is decreased and output speed is increased. Who is right?
    A. A only
    B. B only
    C. Both A and B
    D. Neither A nor B
11. The most typical way to identify a transmission is usually by:
    A. the shape of the oil pan.
    B. the owner guide.
    C. the service manual.
    D. the Vehicle Identification Number (VIN).
12. A vehicle comes in with fluid coming out of the transmission vent. Technician A says that someone may have overfilled it. Technician B says the transmission may have overheated. Who is right?
    A. A only
    B. B only
    C. Both A and B
    D. Neither A nor B

Answers to the Test Questions for the Pretest

1. C, 2. D, 3. D, 4. B, 5. D, 6. D, 7. B, 8. C, 9. B, 10. C, 11. A, 12. C

Explanations to Selected Answers for the Pretest

Question #1
Answer A is wrong.
Answer B is wrong.
Answer C is correct. Both low engine vacuum and poor engine performance can cause delayed shifts.
Answer D is wrong.

Question #3
Answer A is wrong. Replacement gasket sets will supply cast iron hook end seal rings in place of Teflon™ seals.
Answer B is wrong. Teflon™ seals can be installed by hand.
Answer C is wrong.
Answer D is correct.

Question #5
Answer A is wrong. All seals should be replaced.
Answer B is wrong. You should only use automatic transmission fluid (ATF) and vaseline to lubricate parts when assembling a transmission.
Answer C is wrong.
Answer D is correct.

Question #7
Answer A is wrong. Universal joints are only there to allow the power to be redirected at different angles.
Answer B is correct. As the suspension moves, the driveshaft needs to move.
Answer C is wrong.
Answer D is wrong.

Question #9
Answer A is wrong. The converter is bolted to the flexplate but not the rear hub, it is bolted to the front.
Answer B is correct. The flexplate allows the converter to flex.
Answer C is wrong.
Answer D is wrong.

Question #11
Answer A is correct. The shape of the oil pan is the fastest and most common method of identification.
Answer B is wrong. Most owner guides do not say which transmission is in or available in the vehicle.
Answer C is wrong. Although you could identify it through the service manual, it is not the easiest or most common way.
Answer D is wrong. Although you could identify it through the VIN, it is not the easiest or most common way.

Types of Questions

ASE certification tests are often thought of as being tricky. They may seem to be tricky if you do not completely understand what is being asked. The following examples will help you recognize certain types of ASE questions and avoid common errors.

Each test is made up of forty to eighty multiple-choice questions. Multiple-choice questions are an efficient way to test knowledge. To answer them correctly, you must think about each choice as a possibility, and then choose the one that best answers the question. To do this, read each word of the question carefully. Do not assume you know what the question is about until you have finished reading it.

Multiple-Choice Questions

One type of multiple-choice question has three wrong answers and one correct answer. The wrong answers, however, may be almost correct, so be careful not to jump at the first answer that seems to be correct. If all the answers seem to be correct, choose the answer that is the most correct. If you readily know the answer, this kind of question does not present a problem. If you are unsure of the answer, analyze the question and the answers. For example:

Question 1:

A rocker panel is a structural member of which vehicle construction type?

A. Front-wheel drive

B. Pickup truck

C. Unibody

D. Full-frame

Analysis:

This question asks for a specific answer. By carefully reading the question, you will find that it asks for a construction type that uses the rocker panel as a structural part of the vehicle.

Answer A is wrong. Front-wheel drive is not a vehicle construction type.

Answer B is wrong. A pickup truck is not a type of vehicle construction.

Answer C is correct. Unibody design creates structural integrity by welding parts together, such as the rocker panels, but does not require exterior cosmetic panels installed for full strength.

Answer D is wrong. Full frame describes a body-over-frame construction type that relies on the frame assembly for structural integrity.

Therefore, the correct answer is C. If the question was read quickly and the words "construction type" were passed over, answer A may have been selected.

EXCEPT Questions

Another type of question used on ASE tests has answers that are all correct except one. The correct answer for this type of question is the answer that is wrong. The word "EXCEPT" will always be in capital letters. You must identify which of the choices is the wrong answer. If you read quickly through the question, you may overlook what the question is asking and answer the question with the first correct statement. This will make your answer wrong. An example of this type of question and the analysis is as follows:

Question 2:

All of the following are tools for the analysis of structural damage EXCEPT:

A. height gauge.

B. tape measure.

C. dial indicator.

D. tram gauge.

Analysis:

The question really requires you to identify the tool that is not used for analyzing structural damage. All tools given in the choices are used for analyzing structural damage except one. This question presents two basic problems for the test-taker who reads through the question too quickly. It may be possible to read over the word "EXCEPT" in the question or not think about which type of damage analysis would use answer C. In either case, the correct answer may not be selected. To correctly answer this question, you should know what tools are used for the analysis of structural damage. If you cannot immediately recognize the incorrect tool, you should be able to identify it by analyzing the other choices.

Answer A is wrong. A height gauge *may* be used to analyze structural damage.

Answer B is wrong. A tape measure may be used to analyze structural damage.

Answer C is correct. A dial indicator may be used as a damage analysis tool for moving parts, such as wheels, wheel hubs, and axle shafts, but would not be used to measure structural damage.

Answer D is wrong. A tram gauge *is* used to measure structural damage.

Technician A, Technician B Questions

The type of question that is most popularly associated with an ASE test is the "Technician A says... Technician B says... Who is right?" type. In this type of question, you must identify the correct statement or statements. To answer this type of question correctly, you must carefully read each technician's statement and judge it on its own merit to determine if the statement is true.

Typically, this type of question begins with a statement about some analysis or repair procedure. This is followed by two statements about the cause of the problem, proper inspection, identification, or repair choices. You are asked whether the first statement, the second statement, both statements, or neither statement is correct. Analyzing this type of question is a little easier than the other types because there are only two ideas to consider although there are still four choices for an answer.

Technician A... Technician B questions are really double-true-false questions. The best way to analyze this kind of question is to consider each technician's statement separately. Ask yourself, is A true or false? Is B true or false? Then select your answer from the four choices. An important point to remember is that an ASE Technician A... Technician B question will never have Technician A and B directly disagreeing with each other. That is why you must evaluate each statement independently. An example of this type of question and the analysis of it follows.

Question 3:

Structural dimensions are being measured. Technician A says comparing measurements from one side to the other is enough to determine the damage. Technician B says a tram gauge can be used when a tape measure cannot measure in a straight line from point to point. Who is right?

A. A only

B. B only

C. Both A and B

D. Neither A nor B

Analysis:

With some vehicles built asymmetrically, side-to-side measurements are not always equal. The manufacturer's specifications need to be verified with a dimension chart before reaching any conclusions about the structural damage.

Answer A is wrong. Technician A's statement is wrong. A tram gauge would provide a point-to-point measurement when a part, such as a strut tower or air cleaner, interrupts a direct line between the points.

Answer B is correct. Technician B is correct. A tram gauge can be used when a tape measure cannot be used to measure in a straight line from point to point.

Answer C is wrong. Since Technician A is not correct, C cannot be the correct answer.

Answer D is wrong. Since Technician B is correct, D cannot be the correct answer.

Questions with a Figure

About 10 percent of ASE questions will have a figure, as shown in the following example:

Question 4:

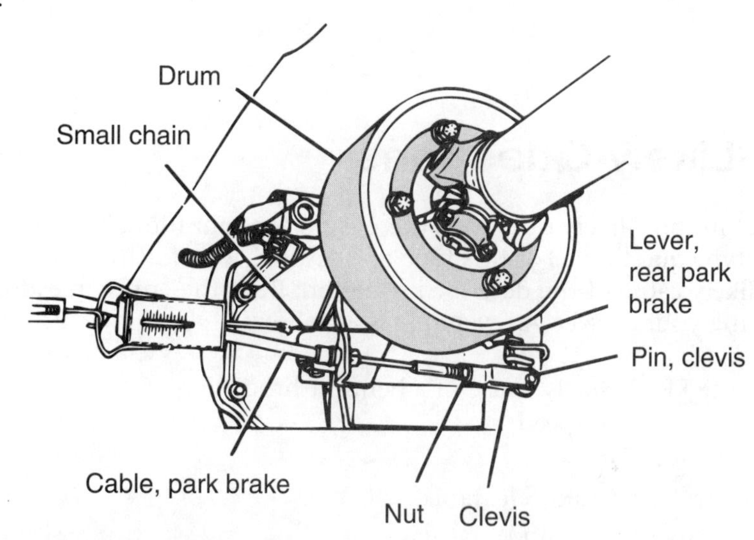

In the measurement shown in the figure above:

A. the driveshaft center support bearing wear is measured.

B. the propeller shaft parking brake is adjusted.

C. the brake shoes must be adjusted before this measurement is taken.

D. the parking brake is released during this measurement.

Analysis:

Answer A is wrong. The driveshaft center support bearing wear is not being measured in this figure.

Answer B is correct. The propeller shaft parking brake adjustment is being performed.

Answer C is wrong. The brake shoes need not be adjusted before adjusting the propeller shaft parking brake.

Answer D is wrong. The parking brake is applied during the propeller shaft parking brake adjustment procedures.

Most-Likely Questions

Most-likely questions are somewhat difficult because only one choice is correct while the other three choices are nearly correct. An example of a most-likely-cause question is as follows:

Question 5:

The most likely cause of reduced turbocharger boost pressure may be a:

A. westgate valve stuck closed.

B. westgate valve stuck open.

C. leaking westgate diaphragm.

D. disconnected westgate linkage.

Analysis:

Answer A is wrong. A westgate valve stuck closed increases turbocharger boost pressure.
Answer B is correct. A westgate valve stuck open decreases turbocharger boost pressure.
Answer C is wrong. A leaking westgate valve diaphragm increases turbocharger boost pressure.
Answer D is wrong. A disconnected westgate valve linkage will increase turbocharger boost pressure.

LEAST-Likely Questions

Notice that in most-likely questions there is no capitalization. This is not so with LEAST-likely type questions. For this type of question, look for the choice that would be the least likely cause of the described situation. Read the entire question carefully before choosing your answer. An example is as follows:

Question 6:

What is the LEAST likely cause of a bent pushrod?

A. Excessive engine speed

B. A sticking valve

C. Excessive valve guide clearance

D. A worn rocker arm stud

Analysis:

Answer A is wrong. Excessive engine speed may cause a bent pushrod.
Answer B is wrong. A sticking valve may cause a bent pushrod.
Answer C is correct. Excessive valve clearance will not generally cause a bent pushrod.
Answer D is wrong. A worn rocker arm stud may cause a bent pushrod.

Summary

There are no four-part multiple-choice ASE questions having "none of the above" or "all of the above" choices. ASE does not use other types of questions, such as fill-in-the-blank, completion, true-false, word-matching, or essay. ASE does not require you to draw diagrams or sketches. If a formula or chart is required to answer a question, it is provided for you. There are no ASE questions that require you to use a pocket calculator.

Testing Time Length

An ASE test session is four hours and fifteen minutes. You may attempt from one to a maximum of four tests in one session. It is recommended, however, that no more than a total of 225 questions be attempted at any test session. This will allow for just over one minute for each question.

Visitors are not permitted at any time. If you wish to leave the test room, for any reason, you must first ask permission. If you finish your test early and wish to leave, you are permitted to do so only during specified dismissal periods.

Monitor Your Progress

You should monitor your progress and set an arbitrary limit to how much time you will need for each question. This should be based on the number of questions you are attempting. It is suggested that you wear a watch because some facilities may not have a clock visible to all areas of the room.

Registration

Test centers are assigned on a first-come, first-served basis. To register for an ASE certification test, you should enroll at least six weeks before the scheduled test date. This should provide sufficient time to assure you a spot in the test center. It should also give you enough time for study in preparation for the test. Test sessions are offered by ASE twice each year, in May and November, at over six hundred sites across the United States. Some tests that relate to emission testing also are given in August in several states.

To register, contact Automotive Service Excellence/American College Testing at:

ASE/ACT
P.O. Box 4007
Iowa City, IA 52243

4 An Overview of the System

Automatic Transmission/Transaxle (Test A2)

The following section includes the task areas and task lists for this test and a written overview of the topics covered in the test.

The task list describes the actual work you should be able to do as a technician that you will be tested on by the ASE. This is your key to the test and you should review this section carefully. We have based our sample test and additional questions upon these tasks, and the overview section will also support your understanding of the task list. ASE advises that the questions on the test may not equal the number of tasks listed; the task lists tell you what ASE expects you to know how to do and be ready to be tested upon.

At the end of each question in the Sample Test and Additional Test Questions sections, a letter and number will be used as a reference back to this section for additional study. Note the following example: **C.13**.

Task List

C. In-Vehicle Transmission/Transaxle Repair (10 Questions)

Task 13 Inspect, replace, and align power train mounts.

Example:

45. A front-wheel-drive vehicle experiences intermittent shifting. Sometimes the transaxle shifts normally but occasionally it misses a shift. Technician A says the manual valve shift linkage may need adjusting. Technician B says the engine or transaxle may be broken. Who is right?
    A. A only
    B. B only
    C. Both A and B
    D. Neither A nor B (C.13)

Analysis:
Question #45
Answer A is wrong. If the manual valve shift linkage were misadjusted, it would cause a problem all of the time, not just sometimes.
Answer B is correct.
Answer C is wrong.
Answer D is wrong.

Task List and Overview

A. General Transmission/Transaxle Diagnosis (24 Questions)

1. Mechanical/Hydraulic Systems (12 Questions)

Task 1 **Listen to driver's concern and road test vehicle to verify mechanical/hydraulic system problems; determine necessary action.**

Before beginning to rebuild or repair a transmission, make sure it has a problem. Your diagnosis should begin with a fluid level check. If the fluid is low, the problem could be fluid leaks. Check the transmission case, oil pan, and cooler lines for evidence of leaks. Check the vacuum line to the vacuum modulator for signs of fluid; the modulator diaphragm may be leaking fluid through to the vacuum line and the engine may be burning it. Low fluid levels can cause low pressures, which will cause slippage between shifts. The condition of the fluid should be checked while checking the fluid level. Examine the fluid carefully.

Task 2 **Diagnose noise and vibration problems; determine necessary action.**

A customer will often complain of a transmission noise which, in reality, is caused by something else in the driveline and not the transmission or torque converter. Bad constant-velocity (CV) or universal (U) joints, wheel bearings, and brakes can generate noises that customers mistake for the transmission and torque converter. The entire driveline should be checked before assuming the noise is transmission related.

Most vibrations are caused by an unbalanced torque converter assembly, a poorly mounted torque converter, or a faulty output shaft. The key to determining the cause of the vibration is to pay particular attention to the vibration in relationship to engine speed. If the vibration changes with a change in engine speed, the cause is probably the output shaft or the driveline connected to it.

Noise problems are also best diagnosed by paying a great deal of attention to the speed and condition at which the noise occurs. The conditions to which to pay the most attention are the operating gear and the load on the driveline. If the noise is engine speed related and is present in all gears including Park and Neutral, the most probable source of the noise is the oil pump because it rotates whenever the engine is running. However, if the noise is engine related and is present in all gears except Park and Neutral, the most probable sources of the noise are those parts that rotate in all gears, such as the drive chain, the input shaft, and the torque converter. Noises that occur only when a particular gear is operated must be related to those components responsible for providing that gear, such as a band or clutch. If the noise is vehicle related, the most probable causes are the output shaft and final drive assembly. Often the exact cause of the noise and vibration can only be identified through careful inspection of a disassembled transmission.

Task 3 **Diagnose unusual fluid usage, type, level, and condition problems; determine necessary action.**

The normal color of automatic transmission fluid (ATF) is pink or red. If the fluid has a dark brownish or blackish color and/or burned odor, the fluid has been overheated. A milky white color indicates that engine coolant has been leaking into the transmission's cooler in the radiator. After checking the ATF fluid level and color, wipe the dipstick on absorbent white paper and look at the stain left by the fluid. Dark particles are normally band and clutch material, and white, silvery metal particles are normally caused by the wearing of the transmission's metal parts. If the dipstick cannot be wiped clean, it is probably covered with varnish, which results from fluid oxidation. Varnish or other heavy deposits indicate the need to change the transmission fluid and filter. Low fluid levels can cause a variety of problems. Air can be drawn in the oil pump's inlet circuit and mixed with the fluid. This will result in aerated fluid, which causes slow pressure buildup and low pressures, which will cause slippage between shifts. Air in the pressure regulator valve will cause a buzzing noise when the valve tries to regulate pump pressure. Excessively high fluid levels can also cause aeration. As the planetary gears rotate in high fluid levels, air can be forced into the fluid. Aerated fluid can foam, overheat, and oxidize. All of these problems can interfere with normal valve, clutch, and servo operation. Foaming may be evident by fluid leakage from the transmission's vent.

Task 4 Perform pressure tests; determine necessary action.

Most transmission problems can be identified without conducting a pressure test; therefore, a pressure test should never be heavily relied on or used to begin your diagnostics. A pressure test has its greatest value when the transmission shifts roughly or when the shift timing is wrong. Both of these problems may be caused by excessive line pressure, which can be verified by a pressure test.

During a road test, observe the starting pressures and the steadiness of the increases that occur with slight increases in load. The amount the pressure drops as the transmission shifts from one gear to another should also be noted. The pressure should not drop more than 15 psi (103 kPa) between shifts. Any pressure reading not within the specifications indicates a problem. Typically, when the fluid pressures are low, there is an internal leak, clogged filter, low oil pump output, or faulty pressure regulator valve. If the fluid pressure increased at the wrong time or the pressure was not high enough, sticking valves or leaking seals are indicated. If the pressure drop between shifts was greater then 15 psi, an internal leak at a servo or clutch seal is indicated.

To maximize the usefulness of a pressure test and to be better able to identify specific problems, begin the test by measuring line pressure. Main line pressure should be checked in all gear ranges and at the three basic engine speeds. If the pressure in all operating gears is within specifications at slow idle, the pump and pressure regulators are working fine. If all pressures are low at slow idle, it is likely that there is a problem in the pump, pressure regulator, filter, or fluid level, or there is an internal pressure leak. To further identify the cause of the problem, check the pressure in the various gears while the engine is at fast idle. If the pressures at fast idle are within specifications, the cause of the problem is normally a worn oil pump; however, the problem may be an internal leak. Internal leaks typically are more evident in a particular gear range because that is when ATF is being sent to a particular device through a particular set of valves and passages. If any of these leak, the pressure will drop when that gear is selected or when the transmission is operating in that gear.

Further diagnostics can be made by observing the pressure change when the engine is operating at wide open throttle (WOT) in each gear range. A clogged oil filter will normally cause a gradual drop at higher engine speeds because the fluid cannot pass through the filter fast enough to meet the needs of the transmission and faster turning pump. If the fluid pressure did not change with the increase in engine speed, a stuck pressure regulator may still allow the pressure to build with an increase in engine speed, but it will not provide the necessary boost pressures. If the pressures are high at slow idle, a faulty pressure regulator or throttle valve problem is indicated. If the fall of the pressures are low at WOT, pull on the throttle valve (TV) cable or disconnect the vacuum hose leading to the vacuum modulator. If this causes the pressure to be in the normal range, the low pressure is caused by a faulty cable or there is a problem in the vacuum modulator or vacuum lines. If the pressures stay below specifications, the most likely causes of the problem are the pump or the control system. If all pressures are high at WOT, compare the readings to those taken at slow idle. If they are high at slow idle and WOT, a faulty pressure regulator or throttle system is indicated. If the pressures are normal at slow idle and high at WOT, the throttle system is faulty. To verify that the low pressures are caused by a weak or worn oil pump, conduct a reverse stall test. If the pressures are low during this test but are normal during all other tests, a weak pump is indicated.

Task 5 Perform stall tests; determine necessary action.

If the torque converter and transmission are functioning properly, the engine will reach a specific speed. If the tachometer indicates a speed above or below specification, a possible problem exists in the transmission or torque converter. If a torque converter is suspected of being faulty, it should be removed and the one-way clutch should be checked on the bench.

If the stall speed is below specifications, a restricted exhaust or slipping stator clutch is indicated. If the stator's one-way clutch is not holding, ATF leaving the turbine of the converter works against the rotation of the impeller and slows down the engine. With both of these problems, the vehicle will exhibit poor acceleration, either because of lack of power from the engine or because there is no torque multiplication occurring in the converter.

If the stall speed is only slightly below normal, the engine is probably not producing enough power and should be diagnosed and repaired. Do not perform the stall test for more than three seconds to prevent overheating the transmission.

If the stall speed is above specification, the bands or clutches in the transmission may be slipping and not holding properly. If the vehicle has poor acceleration but had good results from the stall tests, suspect a seized one-way clutch. Excessively hot ATF in the transmission is a good indication that the clutch is seized. However, other problems can cause these same symptoms, so be careful during diagnosis.

A normal stall test will generate a lot of noise, most of which is normal. However, if any metallic noises are heard during the test, diagnose the source of these noises. Operate the vehicle at low speeds on a hoist with the drive wheels free to rotate. If the noises are still present, the source of the noise is probably the torque converter. The converter should be removed and bench tested for internal interference.

Task 6 Perform lockup converter mechanical and hydraulic system tests; determine necessary action.

Nearly all late-model transmissions are equipped with a lockup torque converter. Most of these lockup converters are controlled by the engine control module or computer. The computer turns on the converter clutch solenoid, which opens a valve and allows fluid pressure to engage the clutch. Care should be taken during diagnostics because poor lockup clutch action can be caused by engine, electrical, clutch, or torque converter problems.

The engagement of the lockup clutch should be smooth. If the clutch prematurely engages or is not being applied by full pressure, a shudder or vibration results from rapid grabbing and slipping of the clutch. The clutch begins to lock up and then slips because it cannot hold the engine's torque and complete the lockup. The capacity of the clutch is determined by the oil pressure applied to the clutch and the condition of the frictional surfaces of the clutch assembly.

If the shudder is only noticeable during the engagement of the clutch, the problem is typically in the converter. When the shudder is only evident after the engagement of the clutch, the cause of the shudder is the engine, transmission, or another component of the driveline. You can identify the source of the shudder by first disconnecting the torque converter clutch solenoid or valve, then road testing the vehicle. If the shudder is no longer present, the source of the shudder is the torque converter clutch assembly. If the shudder is caused by the clutch, the converter must be replaced to correct the problem.

When clutch apply pressure is low and the clutch cannot lock firmly, shudder will occur. This may be caused by a faulty clutch solenoid valve or its return spring. The valve is normally held in position by a coil type return spring. If the spring loses tension, the clutch will be able to prematurely engage. Because insufficient pressure is available to hold the clutch, shudder occurs as the clutch begins to grab and then slips. If the solenoid valve and/or return spring is faulty, it should be replaced, as should the torque converter.

All testing of torque converter clutch (TCC) controls should begin with a basic inspection of the engine and transmission. Apparent transmission and torque converter problems are often caused by engine mechanical problems, broken or incorrectly connected vacuum hoses, wrong ignition timing, or wrong idle speed adjustments. Diagnosis of electrical TCC controls is not as hard as some would lead you to believe. In fact, it becomes easier as you understand how electricity works and how the component or system you are diagnosing works. TCC electrical problems are typically diagnosed through visual inspection, computer code retrieval, and electrical checks of specific circuits.

Task 7 Diagnose mechanical and vacuum control systems; determine necessary action.

Mechanical and/or vacuum control can contribute to shifting problems. The condition and adjustment of the various linkages and cables should be checked whenever there is a shifting problem. If upshifts do not occur at the correct speeds or do not occur at all, the problem may be a faulty vacuum modulator or the vacuum supply line to the modulator. Normal vacuum readings indicate that there are no vacuum leaks in the line and it can be assumed that the engine's condition is satisfactory. If you find transmission fluid in the vacuum line to the modulator, the vacuum diaphragm in the modulator is leaking and you should replace the modulator. To verify this, apply a vacuum to the valve with a hand-held pump. The valve will not hold vacuum if the diaphragm is leaking. If the problem seems to be modulator-related but the vacuum source, vacuum lines, and vacuum modulator all check out fine, the modulator may need to be adjusted.

The modulator must be removed from the transmission in order to adjust it. Some modulators are screwed into the transmission case while others are retained by a clamp and cap screw. While removing the modulator assembly, take care not to lose the actuator pin, which may drop out as the valve is removed. Use a hand-held vacuum pump and the required special gauge pins to bench test and adjust the modulator. Always follow the manufacturer's guidelines when making this adjustment.

Air pressure tests are also performed during disassembly to locate leaking seals and during reassembly to check the operation of the clutches and servos. An air pressure test is conducted by applying clean, moisture-free air at pressure to the case holes that lead to the servo and clutch apply passages. You should clearly hear the action of the holding devices. If a hissing noise is heard, a seal is probably leaking in that circuit. If you cannot hear the action of the servo or if it does not react immediately to the introduction of the apply air, something is making it stick. Repair or replace the apply devices if they do not operate normally.

Air pressure can normally be directed to the following circuits through the appropriate hole for each: front clutch, rear clutch, kickdown servo, low servo, and reverse servo. Some manufacturers recommend the use of a specially drilled plate, which is bolted to the transmission case. This plate not only clearly identifies which passages to test but also seals off the other passages. Air is applied directly through the holes on the plate.

2. Electronic Systems (12 Questions)

Task 1 Listen to driver's concern and road test vehicle to verify electronic system problems; determine necessary action.

Some transmissions are only partially controlled; that is, only the engagement of the third to fourth shifting is electronically controlled. Other models feature electronic shifting into all gears plus electronic control of the TCC.

Critical to proper diagnosis of the electronic automatic transmission (EAT) and TCC control system is a road test. The road test should be conducted in the same way as one for a nonelectronic transmission except that a scan tool should also be connected to the circuit to monitor engine and transmission operation. All pressure changes should be noted. The various computer inputs should also be monitored and the readings recorded for future reference. Some scan tools are capable of printing out a report. If the scanner does not have that function, you should record the information after each gear or condition change.

Task 2 Perform pressure tests; determine necessary action.

If you cannot identify the cause of a transmission problem from your inspection or road test, a pressure test should be conducted. This test measures the fluid pressure of the different transmission circuits during various operating gears and gear selector positions. The number of hydraulic circuits that can be tested varies with different makes and models.

Most transmission problems can be identified without conducting a pressure test; therefore, a pressure test should never be heavily relied on to make your diagnosis. A pressure test has its greatest value when the transmission shifts roughly or when the shift timing is wrong. Both of these problems may be caused by excessive line pressure, which can be verified by a pressure test. The results from a pressure test may have little value when diagnosing some problems. If a transmission does not operate in a particular gear but operates fine in other gears, a pressure test will not identify the problem. If there is enough oil pressure to operate the transmission in the other gears, then there is certainly enough pressure to operate in the malfunctioning one. When there is a specific failure or slippage in any gear, the cause of the problem is identified more easily by a complete visual inspection, a road test, and the use of logic.

Task 3 Perform lockup converter electronic system tests; determine necessary action.

All testing of TCC components should begin with a basic inspection of wires and hoses; look for burnt spots, bare wires, and damaged or pinched wires. Make sure the harness to the electronic control unit has a tight and clean connection. Also check the source voltage at the battery before beginning any detailed test on an electronic control system. If the voltage is too low or too high, the electronic system cannot function properly.

On early TCC equipped vehicles, lockup was controlled hydraulically. A switch valve was controlled by two other valves. The lockup valve responds to governor pressure and prevents lockup at speeds below 40 mph (64 km/h). The fail-safe valve responds to throttle pressure and permits lockup in high gear only. Care should be taken during diagnostics because poor lockup clutch action can be caused by engine, electrical, clutch, or torque converter problems.

Before the lockup clutch is applied, the vehicle must be traveling at or above a certain speed. The vehicle speed sensor (VSS) sends this speed information to the computer. The converter should not be able to engage the lockup clutch when the engine is cold, a coolant temperature sensor (CTS) provides the computer with information on temperature. During sudden deceleration or acceleration, the lockup clutch should be disengaged. One of the sensors used to tell the control computer these driving modes is the throttle position sensor (TPS). Some transmissions use a third or fourth gear switch to signal the computer when the transmission is in those gears. A brake switch is used in some circuits to disengage the clutch when the brakes are applied. These key sensors, the VSS, CTS, TPS, third/fourth gear, and brake switch, should be inspected as part of your diagnosis.

Task 4 Diagnose electronic transmission control systems using appropriate test equipment; determine necessary action.

Many switches are used as inputs or control devices for electronic automatic transmissions (EAT). Most of the switches are either mechanically or hydraulically controlled. The operation of these switches can be checked with an ohmmeter. The manual lever position switch provides information to the computer on the range of the transmission. Because this switch is either open or closed, depending on its position it can be checked with an ohmmeter. Pressure switches either complete or open an electrical circuit. These switches are either grounding switches or they connect or disconnect two wires. In early TCC controls, a grounding type switch was tapped into the governor pressure to ground the TCC solenoid. Other grounding type switches are used to tell when a gear is operating or to operate the backup lights. Normally, open switches will have no continuity across the terminals until oil pressure is applied to them. By using air pressure, you can easily tell if a switch is working or if it has a leak.

Another type of switch is a potentiometer. Rather than opening or closing a circuit, a potentiometer controls the circuit by varying its resistance in response to something. A TPS is a potentiometer. It sends very low voltage to the computer when the throttle is closed and increases the voltage as the throttle is opened. Most TPS receive a reference voltage of five volts. A TPS can be checked with an ohmmeter or voltmeter.

Vehicle speed sensors provide road speed to the computer. There are two types of speed sensors: alternating current (AC) voltage generator and reed style sensors. Both of these rely on magnetic principals. The reed style can be checked by rotating the output shaft one revolution; an ohmmeter should show an open and closed circuit within one revolution. AC voltage generators generate voltage each time a tooth passes through the magnetic field; by counting the number of teeth on the output shaft, you can determine how many pulses per one revolution you will measure with a voltmeter set to AC volts.

Temperature sensors are designed to change resistance with changes in temperature. A temperature sensor is based on a thermistor. Some thermistors can increase with an increase in temperature; others decrease as temperature increases. These sensors can be checked with an ohmmeter. Compare the actual temperature and resistance reading to that of charts commonly found in the service manual.

Task 5 Verify proper operation of starting and charging systems; check battery connections and vehicle grounds.

Before diagnosing electrical or electronic problems, a check of the vehicle's charging/starting system should be performed. A hydrometer tests the specific gravity of the battery electrolyte. A fully charged battery should have a specific gravity of 1.265. A battery load capacity tester is used to test the battery capacity. During this test, the battery is discharged at one-half the cold cranking amperes for 15 seconds, and the voltage is recorded at the end of this time. A satisfactory battery at 70°F (21°C) has 9.6 volts at the end of the capacity test.

Most starters, relays, solenoids, and circuits may be tested by measuring the voltage drop across the various components and wires with the starting motor cranking the engine. If the voltage drop across any component or connecting wire exceeds the manufacturer's specifications, replace the component or wire.

B. Transmission/Transaxle Maintenance and Adjustment (5 Questions)

Task 1 Inspect, adjust, and replace manual valve shift linkage and transmission range sensor/switch.

A worn or misadjusted gear selection linkage will affect transmission operation. The transmission's manual shift valve must completely engage the selector gear. Partial manual shift valve engagement will not allow the proper amount of fluid pressure to reach the rest of the valve body. If the linkage is misadjusted, poor gear engagement, slipping, and excessive wear can result. The gear selector linkage should be adjusted so the manual shift valve detent position in the transmission matches the selector detent and position indicator.

To check the adjustment of the linkage, move the shift lever from the Park position to the lowest Drive gear. Detents should be felt at each of these positions. If the detent cannot be felt in any of these positions, the linkage needs to be adjusted. After adjusting any type of shift linkage, recheck it for detents throughout its range. As a safety measure, make sure a positive detent is felt when the shift lever is placed into the Park position. If you are unable to make an adjustment, the levers and grommets may be badly worn or damaged, and should be replaced. When it is necessary to disassemble the linkage from the levers, the grommets used to retain the cable or rod should be replaced. Use a prying tool to force the cable or rod from the grommet, and then cut the old grommet. Pliers can be used to snap the new grommets into the levers and the cable or rod into the levers.

Task 2 Inspect, adjust, and replace cables or linkages for throttle valve (TV), kickdown, and accelerator pedal.

The throttle valve cable connects the movement of the throttle pedal movement to the TV in the transmission's valve body. On some transmissions, the throttle linkage may control both the downshift valve and the throttle valve. Others use a vacuum modulator to control the TV and throttle linkage to control the downshift valve. Later model transmissions may not have a throttle cable. Instead, they rely on electronic sensors and switches to monitor engine load and the throttle's opening. The action of the throttle valve produces throttle pressure. Throttle pressure is used as an indication of engine load and influences the speed at which automatic shifts will take place.

A misadjusted TV linkage may also result in a throttle pressure that is too low in relation to the amount the throttle plates are open, causing early upshifts. Too high of a throttle pressure can cause harsh and delayed upshifts, and part throttle and WOT downshifts will occur earlier than normal. An adjustment as small as a half-turn can make a big difference in shift timing and feel.

Most late-model transmissions are not equipped with downshift linkages; rather they use a kickdown switch typically located at the upper post of the throttle pedal. Movement of the throttle pedal to the wide-open position signals to the transmission that the driver desires a forced downshift. To check the switch, fully depress the throttle pedal and listen for a click that should be heard just before the pedal reaches its travel stop. If the click is not heard, loosen the locknut and extend the switch until the pedal lever makes contact with the switch. If the pedal contacts the switch too early, the transmission may downshift during part throttle operation. If you hear the click but the transmission still does not downshift, use an ohmmeter to check the switch. An open switch will prevent forced downshifts whereas a shorted switch can cause upshift problems.

Task 3 Adjust bands, where applicable.

On some transmissions, slippage during shifting can be corrected by adjusting the holding bands. To help identify if a band adjustment will correct the problem, refer to the written results of your road test. Compare your results with the clutch and band application chart in your service manual. If slippage occurs when there is a gear change that requires the holding by a band, the problem may be corrected by tightening the band. On some vehicles, the bands can be adjusted externally with a torque wrench. On others, the transmission fluid must be drained and the oil pan removed.

Task 4 Replace fluid and filter(s).

The transmission's fluid and filter should be changed whenever there is an indication of oxidation or contamination. Periodic fluid and filter changes are also part of the preventative program for most vehicles. The frequency of this service depends on the conditions under which the transmission is normally operated. Severe usage requires that the fluid and filter be changed more often.

Change the fluid only when the engine and transmission are at normal operating temperatures. On most transmissions, you must remove the oil pan to drain the fluid.

Some transmission pans on recent vehicles include a drain plug. A filter or screen is normally attached to the bottom of the valve body. Filters are made of paper or fabric and are held in place by screws, clips, or bolts. Filters should be replaced, not cleaned. After draining, check the bottom of the pan for deposits and metal particles. Slight contamination, or blackish deposits from clutches and bands, is normal. Other contaminants should be a concern. Steel particles indicate severe internal transmission wear or damage. If the metal particles are aluminum, they may be part of the torque converter stator. Some torque converters us phenolic plastic stators; therefore, metal particles found in these transmissions must be from the transmission itself. Filters are always replaced, whereas screens are cleaned. Screens are removed in the same way as filters. Clean a screen with fresh solvent and a stiff brush.

C. In-Vehicle Transmission/Transaxle Repair (10 Questions)

Task 1 Inspect, adjust, and replace vacuum modulator, valve, lines, and hoses.

Diagnosing the vacuum modulator begins with checking the vacuum at the line or hose connected to the modulator. The modulator should be receiving engine manifold vacuum. If it does, and there are no vacuum leaks in the line to the modulator, check the modulator itself for leaks with a hand-held vacuum pump. If transmission fluid is found when you disconnect the line at the modulator, the vacuum diaphragm in the modulator is leaking and the modulator should be replaced. If the vacuum source, vacuum lines, and vacuum modulator are in good condition but shifts indicate a vacuum modulator problem, the modulator may need adjusting. Most modulators must be removed to be adjusted; however, there are some that have an external adjustment. This adjustment allows for fine-tuning of modulator action.

Task 2 Inspect, adjust, repair, and replace governor cover, seals, sleeve/bore, valve, weight, springs, retainers, and gear.

If the pressure tests suggested that there was a governor problem, it should be removed, disassembled, cleaned, and inspected. Some governors are mounted internally, and the transmission must be removed to service the governor. Other governors can be serviced by removing the extension housing or oil pan, or by detaching an external retaining clamp and then removing the unit.

Improper shift points are typically caused by a faulty governor or governor drive gear system. However, some electronically controlled transmissions do not rely on the hydraulic signals from a governor; rather, they rely on electrical signals from sensors. Sensors such as speed and load sensors signal to the transmission's computer when gears should be shifted. Faulty electrical components and/or loose connections can also cause improper shift points.

Task 3 Inspect and replace external seals and gaskets.

Continue your diagnostics by conducting a quick and careful visual inspection. Check all drive train parts for looseness and leaks. If the transmission fluid was low or there was no fluid, raise the vehicle and carefully inspect for signs of leakage. Leaks are often caused be defective gaskets or seals. Common sources for leaks are the oil pan seal, rear cover, final drive cover (on transaxles), extension housings, speedometer drive gear assembly, and electrical switches mounted into the housing. The housing itself may have a porosity problem, allowing fluid to seep through the metal. Case porosity may be repaired using an epoxy sealer.

A common cause of fluid leakage is the seal of the oil pan to the transmission housing. If there are signs of leakage around the rim of the pan, retorquing the pan bolts may correct the problem. If tightening the pan does not correct the problem, the pan must be removed and a new gasket installed. Make sure the sealing surface of the pan's rim is flat and capable of providing a seal before reinstalling it.

An oil leak at the speedometer cable can be corrected by replacing the O-ring seal. While replacing the seal, inspect the drive gear for chips and missing teeth. Always lubricate the O-ring and gear prior to installation.

Task 4 Inspect and repair or replace extension housing; replace bushing.

An oil leak stemming from the mating surface of the extension housing and the transmission case may be caused by loose bolts. To correct this problem, tighten the bolts to the specified torque. Also check for signs of leakage at the rear of the extension housing. Fluid leaks from the seal of the extension housing can be corrected with the transmission in the vehicle. Often, the cause for the leaks is a worn extension housing bushing, which supports the sliding yoke of the driveshaft. When the driveshaft is installed, the clearance between the sliding yoke and bushing should be minimal. If the clearance is satisfactory, a new oil seal will correct the leak. If the clearance is excessive, the repair requires that a new bushing be installed. If the seal is faulty, the transmission vent should be checked for blockage.

Task 5 Check condition of engine cooling system; inspect, test, flush, and replace transmission cooler, lines, and fittings.

Vehicles equipped with an automatic transmission can have an internal or external transmission cooler, or both. The basic operation of either type of cooler is that of a heat exchanger. Heat from the fluid is transferred to something else, such as liquid or air. Hot ATF is sent from the transmission to the cooler, where it has its heat removed, and then the cooled ATF returns to the transmission.

Internal coolers are located inside the engine's radiator. Heated ATF travels from the torque converter to a connection at the radiator. Inside the radiator is a small internal cooler, which is sealed from the liquid in the radiator. ATF flows through this cooler and its heat is transferred to the liquid in the radiator. The ATF then flows out of a radiator connection, back to the transmission.

External coolers are mounted outside the engine's radiator, normally just in front of it. Air flowing through the cooler removes heat from the fluid before it is returned to the transmission.

The engine's cooling system is the key to efficient transmission fluid cooling. If anything affects engine cooling, it will also affect ATF cooling. The engine's cooling system should be carefully inspected whenever there is evidence of ATF overheating or a transmission cooling problem. If the problem is the transmission cooler, examine it for signs of leakage. Check the dipstick for evidence of coolant mixing with the ATF. Milky fluid indicates that engine coolant is leaking into and mixing with the ATF because of a leak in the cooler. At times, the presence of ATF in the radiator will be noticeable when the radiator cap is removed, because ATF will tend to float to the top of the coolant. A leaking transmission cooler core can be verified with a leak test.

External cooler leaks result in traces of ATF or film buildup of ATF around the source of the leak. It usually takes a little time to determine the source of the leakage. However, internal coolers present more difficulties and a leak test should be conducted.

The inability of the transmission fluid to cool can be caused by a plugged or restricted fluid cooling system. If the fluid cannot circulate, it cannot cool. A tube type transmission cooler can be cleaned by using solvent or mineral spirits and compressed air. A fin type cooler, however, cannot be cleaned in the same way. Therefore, normal procedures include replacing the radiator (which includes the cooler). In both cases, the cooler line should be flushed. With the engine idling, the flow through the cooler should be one quart in 20 seconds.

Check the condition of the cooler lines from their beginning to their end. A line that has been accidentally damaged while the transmission has been serviced will reduce oil flow through the cooler and shorten the life of the transmission. If the steel cooler lines need to be replaced, use only double-wrapped and brazed steel tubing. Never use copper or aluminum tubing to replace steel tubing. The steel tubing can be double flared and installed with the correct fittings.

Task 6 Inspect and replace speedometer/speed sensor drive gear, driven gear, and retainers.

The vehicle's speedometer can be purely electronic, which requires no mechanical hookup to the transmission, or it can be driven off the output shaft via a cable. Most oil leaks at the speedometer cable can be corrected by replacing the O-ring seal.

Speedometer gear problems normally result in an inoperative speedometer. However, this problem can also be caused by a faulty cable, drive gear, or speedometer. A damaged drive gear can cause the driven gear to fail, and both should be carefully inspected during a transmission overhaul. On some transmissions, the speedometer drive gear is a set of gear teeth machined into the output shaft. Inspect this gear. If teeth are slightly rough, they can be cleaned up and smoothened with a file. If the gear is severely damaged, the entire output shaft must be replaced. Transmissions have a drive gear that is splined to the output shaft, held in place by a clip, or driven and retained by a ball that fits into a depression in the shaft. If a clip is used, it should be carefully inspected for cracks or other damage.

The drive gear can be removed and replaced if necessary. The driven gear is normally attached to the transmission end of the speedometer cable. If the drive gear is damaged, it is likely that the driven gear will also be damaged. Most driven gears are made of plastic to reduce speedometer noise; they are weaker than most drive gears. Always check the retainer of the driven gear on the speedometer cable.

Task 7 Inspect valve body mating surfaces, bores, valves, springs, sleeves, retainers, brackets, check balls, screens, spacers, and gaskets; replace as necessary.

If the pressure test indicates there is a problem associated with the valves in the valve body, a thorough disassembly, a cleaning in fresh solvent, a careful inspection, and the freeing up and polishing of the valves may correct the problem. Sticking valves and sluggish valve movements are caused by poor maintenance, the use of the wrong type of fluid, and/or overheating the transmission. The valve body of most transmissions can be serviced when the transmission is in the vehicle, but is typically serviced when the transmission has been removed for other repairs.

After all of the valves and springs have been removed from the valve body, soak the valve body and separator plates in mineral spirits for a few minutes. Some rebuild shops soak the valve body and its associated parts in carburetor cleaner, and then wash off the parts with water. Thoroughly clean all parts and make sure all passages within the valve body are clear and free of debris. Carefully blow-dry each part individually with dry compressed air. Never wipe the parts of a valve body with a rag or paper towel. Lint from both will collect in the valve body passages and cause shifting problems.

Check the separator plate for scratches or other damage. Scratches or score marks can cause oil to bypass correct oil passages and result in system malfunction. If the plate is defective in any way, it must be replaced. Check the oil passages in the upper and lower valve bodies for varnish deposits, scratches, or other damage that could restrict the movement of the valves. Check all of the threaded and related bolts and screw for damaged threads, and replace as needed.

Examine each valve for nicks, burrs, and scratches. Make sure that each valve fits properly into its respective bore. To do this, hold the valve body vertically and install an unlubricated valve into its bore. Let the valve fall of its own weight into the valve body until the valve stops. Then place your finger over the valve bore and turn the valve body over. The valve should again drop by its own weight. If the valve moves freely under these conditions, it will operate freely with fluid pressure.

If a valve cannot move freely within its bore, it may have small burrs or nicks. These flaws can and should be removed. To do this, never use sandpaper or a file; rather, use products such as an Arkansas stone or crocus cloth, which are designed to polish the surface without removing metal from the valve. Sandpaper and emery cloth will remove metal, as well as scratch and leave a rough surface. Normally, valve bodies are replaced if they have damaged bores.

After polishing, the valve must be thoroughly cleaned to remove all of the cleaning and abrasive materials. After the valve has been recleaned, it should be tested in its bore again. If the valve cannot be cleaned well enough to move freely in its bore, the valve body should be replaced. Individual valve body parts are not available. Individual valves are lapped to a particular valve body, and, therefore, if any parts need to be replaced, the entire valve body must be replaced.

Although it is desirable to have the valves move freely in their bores, excessive wear is also a problem. There should never be more than 0.001 inch (0.025 mm) clearance between the valve and its bore. If the bore or piston is worn, the entire valve body needs to be replaced.

With a straightedge and feeler gauge, check the flatness of the valve body's sealing surface. If it is warped, it can be flat filed.

Task 8 Check and adjust valve body bolt torque.

Before beginning to reassemble a valve body, check the new valve body gasket to make sure it is the correct one by laying it over the separator plate and holding it up to the light. No oil holes should be blocked. Then, install the bolts to hold the valve body sections together and the valve body to the case. Tighten the bolts to the torque specifications to prevent warpage and possible leaks. Overtorquing can also cause the bores to distort, which would not allow the valves to move freely once the valve body is tightened to the transmission case.

Task 9 Inspect servo bore, piston, seals, pin, spring, and retainers; repair or replace as necessary.

On some transmissions, the servo and accumulator assemblies are serviceable with the transmission in the vehicle. Others require the complete disassembly of the transmission.

Before disassembling a servo or any other components, carefully inspect the area to determine the exact cause of the leakage. Do this before cleaning the area around the seal. Look at the path of the fluid leakage and identify other possible sources. These sources could be worn gaskets, loose bolts, cracked housings, or loose line connections.

Inspect the outside of the seal. If it is wet, determine if the oil is leaking out or if it is merely lubricating film oil. When removing the servo, continue to look for the cause of the leak. Check both inner and outer parts of the seal for wet oil, which means leakage. When removing the seal, inspect the sealing surface, or lips, before washing. Look for unusual wear, warping, cuts and gouges, or particles embedded in the seal.

Band servos and accumulators are basically pistons with seals in a bore that are held in position by springs and retaining snaprings. Remove the retaining rings and pull the assembly from the bore for cleaning. Check the condition of the piston and springs. Cast iron seal rings may not need replacement, but rubber and elastomer seals should always be replaced.

A servo's piston, spring, piston rod, and guide should be cleaned and dried. Check the servo piston for cracks, burrs, scores, and wear. Servo pistons may be made of either aluminum or steel. Aluminum pistons should be carefully checked for cracks and for their fit on the guide pins. Cracked pistons will allow for a pressure loss and pistons that are loose on the guide pin may allow the piston to bind in its bore. Whether it is a steel or aluminum piston, the seal groove should be free of nicks or any imperfection that might pinch or bind the seal. Clean up any problems with a small file or scraper. A side clearance of 0.003 to 0.005 inch (0.076 to 0.127 mm) is required.

Task 10 Inspect accumulator bore, piston, seals, springs, and retainers; repair or replace as necessary.

Begin the disassembly of an accumulator by removing the accumulator plate snapring. After removing the accumulator plate, remove the spring and accumulator pistons. If rubber seal rings are installed on the piston, replace them whenever you are servicing the accumulator. Lubricate the new accumulator piston ring and carefully install it on the piston. Lubricate the accumulator cylinder walls and install the accumulator piston and spring. Then reinstall the accumulator plate and retaining snapring.

Many accumulator pistons can be installed upside down. This results in free travel of the piston or too much compression of the accumulator spring. Note the direction of installation of the piston during the teardown process, because you will not always find a good picture when you need to reinstall the accumulator. Because the movement of the accumulator has an effect on shift feel, correct installation is critical. It is quite common for manufacturers to mate servo piston assemblies with accumulators. This takes up less space in the transmission case and, because they have the same basic shape, can reduce some of the machining during manufacturing.

Task 11 Inspect and replace parking pawl, shaft, spring, and retainer.

The parking pawl can be inspected after the transmission is disassembled or, on some transmissions, while the transmission is still in the vehicle. Examine the engagement lug on the pawl, making sure it is not rounded off. If the lug is worn, it will allow the pawl to possibly slip out or not fully engage in the parking gear. Because most parking pawls pivot on a pin, this also needs to be checked to make sure there is no excessive looseness at this point. The spring that pulls the pawl away from the parking gear must also be checked to make sure it will remain in that position during operation.

The pushrod or operating shaft must provide the correct amount of travel to engage the pawl to the gear. Make sure the shaft is not bent or that the pivot holes in the internal shift linkages are not worn or oblong.

Any components found unsuitable should be replaced. It should be noted that the components that make up the parking lock system are the only parts holding the vehicle in place when parked. If they do not function correctly, the vehicle may roll or even drop into reverse when the engine is running, causing an accident or injury. Replace any questionable or damaged parts.

Task 12 Inspect, test, adjust, repair, or replace electrical/electronic components, including: computers, solenoids, sensors, relays, fuses, terminals, connectors, switches, and harnesses.

Some electronic transmissions are only partially controlled. Only the engagement of the converter and third to fourth shifting is electronically controlled. Other models feature electronic shifting into all gears, plus electronic control of the TCC.

The controls of an electronic automatic transmission (EAT) direct the hydraulic flow through the use of solenoid valves. When it is used to control TCC operation, the solenoid opens a hydraulic circuit to the TCC spool valve, causing the spool valve to move and direct mainline pressure to apply the clutch. Electronically controlled shifting is accomplished in much the same way. Shifting occurs when a solenoid is either turned on or turned off. At least two shift solenoids are incorporated into the system, and shifting takes place by controlling the solenoids. The desired gear is put into operation through a combination of on and off solenoids.

Several sensors and switches are used to inform the control computer of the current operating conditions. Most of these sensors are also used to calibrate engine performance. The computer then determines the appropriate shift time for maximum efficiency and best feel. The shift solenoids are controlled by the computer, which either supplies power to the solenoids or supplies a ground circuit. The techniques for diagnosing electronic transmissions are basically the same techniques used to diagnose TCC systems.

Although EATs are rather reliable, they have introduced new problems for the automatic transmission technician. Some of the common problems that affect shift timing and quality, as well as the timing and quality of TCC engagement, are wrong battery voltage, a blown fuse, poor connections, a defective throttle position sensor (TPS) or vehicle speed sensor (VSS), defective solenoids, crossed wires to the solenoid or sensors, corrosion at an electrical terminal, or the faulty installation of some accessory, such as a cellular telephone.

Improper shift points can be caused by electrical circuit problems, faulty electrical components, or bad connectors, as well as a defective governor or governor drive gear assembly. Some EATs do not rely on the hydraulic signals from a governor; rather they rely on the electrical signals from electrical sensors to determine shift timing.

Computer-controlled transmissions often start off in the wrong gear. This can happen due to either internal transmission problems or external control system problems. Internal transmission problems can be faulty solenoids or stuck valves. External problems can be the result of a complete loss of power or ground to the control circuit or a fail-safe protection strategy initiated by the computer to protect itself or the transmission from an observed problem. Typically, the default gear is simply the gear that is applied when the shift solenoids are off, usually second gear and reverse.

A visual inspection of the transmission and the electrical system should include a careful check of all electrical wires and connectors for damage, looseness, and corrosion. Loose connections, even when clean, usually only make intermittent contact. They will also corrode and collect foreign material, which can prevent contact altogether. Use an ohmmeter to check the continuity through a connector suspected of being faulty. Check all ground straps to the frame or engine block. This part of your inspection is especially important for electronically controlled transmissions that have a lockup torque converter. Check the fuse or fuses to the control module. To accurately check a fuse, either test it for continuity with an ohmmeter or check each side of the fuse for open power when the circuit is activated.

Task 13 Inspect, replace, and align powertrain mounts.

The engine and transmission mounts on front-wheel-drive (FWD) cars are important to the operation of the transaxle. Any engine movement may affect the length of the shift and throttle cables and may affect the engagement of the gears. Delayed or missed shifts may be the result of linkage changes as the engine pivots on its mounts. Problems with transmission mounts may also affect the operation of a rear-wheel-drive (RWD) vehicle, but this type of problem will be less detrimental than the same type of problem on a FWD vehicle. Many shifting and vibration problems can be caused by worn, loose, or broken engine and transmission mounts. Visually inspect the mounts for looseness and cracks. To get a better look at the condition of the mounts, pull up and push down on the transaxle case while watching the mount. If the mount's rubber separates from the metal plate or if the case moves up but not down, replace the mount. If there is movement between the metal plate and its attaching point on the frame, tighten the attaching bolts to the appropriate torque. Then, from the driver's seat, apply the foot brake and start the engine. Put the transmission into a gear and gradually increase the engine speed to about 1,500–2,000 rpm. Watch the torque reaction of the engine on its mounts. If the engine's reaction to the torque appears to be excessive, broken or worn drive train mounts may be the cause. If it is necessary to replace the transaxle mount, make sure you follow the manufacturer's recommendations for maintaining the alignment of the driveline. Failure to do this may result in poor gear shifting, vibrations, and/or broken cables. Some manufacturers recommend that a holding fixture or special bolt be used to keep the unit in its proper location.

When removing the transaxle mount, begin by disconnecting the battery's negative cable. Disconnect any electrical connectors that may be located around the mount. It may be necessary to move some accessories, such as the horn, in order to service the mount without damaging some other assembly. Be sure to label any wires you remove to facilitate reassembly.

Install the engine support fixture and attach it to an engine hoist. Lift the engine just enough to take the pressure off of the mounts. Remove the bolts attaching the transaxle mount to the frame and the mounting bracket, and then remove the mount.

To install the new mount, position the transaxle mount in its correct location on the frame and tighten its attaching bolts to the proper torque. Install the bolts that attach the mount to the transaxle bracket. Prior to tightening these bolts, check the alignment of the mount. Once you have confirmed that the alignment is correct, tighten all loosened bolts to their specified torque. Remove the engine hoist fixture from the engine and reinstall all accessories and wires that may have been removed earlier.

D. Off-Vehicle Transmission/Transaxle Repair (11 Questions)

1. Removal, Disassembly, and Assembly (3 Questions)

Task 1 Remove and replace transmission/transaxle; inspect engine core plugs, transmission dowel pins, and dowel pin holes.

Removing the transmission from a RWD vehicle is generally more straightforward than removing one from a FWD model, as there is typically one crossmember, one driveshaft, and easy access to cables, wiring, cooler lines, and bell housing bolts. Transmissions in FWD cars, because of their limited space, can be more difficult to remove as you may need to disassemble or remove large assemblies such as the engine cradle, suspension components, brake components, splash shields, or other pieces that would not usually affect RWD transmission removal.

On RWD vehicles, raise the vehicle, and drain the transmission fluid. Mark the driveshaft at the rear axle before disconnecting it to avoid runout-related vibrations. Remove the driveshaft.

On FWD vehicles, attach a support fixture to the engine, raise the vehicle, and drain the fluid. Remove the front wheels, and follow the service manual to remove the front axles.

Disconnect manual linkages, vacuum hoses, electrical connections, speedometer drives, and control cables. The inspection cover between the transmission and the engine should be removed next. Mark the position of the converter to the flexplate to help maintain balance or runout. It will be necessary to rotate the crankshaft to remove the converter bolts. This can be done by using a long ratchet and socket on the crankshaft bolt or by using a flywheel turning tool if space permits.

Position a transmission jack before removing any crossmember or bell housing bolts. The use of a transmission jack also allows for easier access to parts hidden by crossmembers or hidden in the space between the transmission and the vehicle's floor pan. With a transmission jack supporting the transmission, remove the crossmember or transmission mounts. Remove the starter and bell housing bolts.

Now pull the transmission away from the engine. It may be necessary to use a pry bar between the transmission and engine block to separate the two units. Make sure the converter comes out with the transmission. This prevents bending the input shaft, damaging the oil pump, or distorting the drive hub. After separating the transmission from the engine, retain the torque converter in the bell housing. This can be done simply by bolting a small combination wrench to a bell housing bolt hole across the outer edge of the converter.

Be sure to check all bell housing bolt holes and dowel pins. Cracks around the bolt holes indicate that the case bolts were tightened with the case out of alignment with the engine block. The case should be replaced if the following problems are present: broken worm tracking, cracked case at the oil pump to case flange, a cracked case at the clutch housing pressure cavity, or ears broken off the case. It is not possible to determine if a repair will hold. A transmission case is very thin, and welding may distort the case.

If any of the bolts that were removed during disassembly have aluminum on the threads, the thread bore is damaged and should be repaired. Thread repair entails installing a thread insert, which serves as new threads for the bolt, or retapping the bore. After the threads have been repaired, make sure you thoroughly clean the case.

Task 2 Disassemble, clean, and inspect.

Before disassembling the automatic transmission, care should be taken to clean away any dirt, undercoating, grease, or road grime on the outside of the case. This ensures that dirt will not enter the transmission during disassembly. Once the transmission is clean outside, you may begin the disassembly. When cleaning automatic transmission parts, avoid the use of solvents, degreasers, and detergents that can decompose the friction composites used in a transmission. Use compressed air to dry components; do not wipe down parts with a rag. The lint from a rag can easily destroy a transmission after it has been rebuilt.

There are many different methods used to clean automatic transmission parts. Some rebuilding shops use a parts washing machine, which takes a little time to thoroughly clean a transmission case and associated parts. These parts washers use hot water and a special detergent that is sprayed onto the parts as they rotate inside the cleaner. Many rebuilders simply clean the parts in mineral spirits tank, where the parts are brushed and hand cleaned. No matter what type of cleaning procedure is followed, the transmission and parts should be rinsed with water and then air-dried before reassembly.

After the case is clean, remove the torque converter and carefully inspect it for damage. Check the converter hub for grooves caused by hardened seals. Also check the bushing contact area. To remove the converter, slowly rotate it as you pull it from the transmission; have a drain pan handy to catch the fluid. It should come out without binding. This is a good time to check the input shaft splines, stator support splines, and the converter's pump drive hub for any wear or damage. Converters with direct driveshafts should be checked to be sure that no excessive play is present at the drive splines of the shaft or the converter. If any play is found in the converter, the converter or the shaft must be replaced.

Position the transmission to perform an end play check. The transmission end play checks can provide the technician with information about the condition of internal bearings and seals, as well as clues to possible causes of improper operation found during the road test. These measurements will also determine the thickness of the thrust washers during reassembly. The thrust washers' thickness sets the end play of various components. Excessive end play allows clutch drums to move back and forth too much, causing the transmission case to wear. Assembled end play measurements should be between minimum and maximum specifications, but preferably at the low end of the specifications.

Task 3 Assemble after repair.

Before proceeding with the final assembly of all components, it is important to verify that the case, housing, and parts are clean and free from dust, dirt, and foreign matter. Have a tray available with clean ATF for lubricating parts; also have a jar or tube of vaseline for securing washers during installation. Coat all parts with the proper type of ATF. Soak bands and clutches in the fluid for at least 15 minutes before installing them. All new seals and rings should have been installed before beginning final assembly.

Carefully examine all thrust washers and coat them with petroleum jelly before placing them in the housing. Install the thrust ring, piston return spring, thrust washer, and one-way clutch inner race into the case. Align and start the bolts into the inner race from the rear of the case. Tighten the bolts to specifications.

Task 4 Inspect converter flex (drive) plate, converter attaching bolts, converter pilot, and converter pump drive surfaces.

The converter's drive lugs or studs should be carefully inspected. These hold the converter firmly to the flexplate and ensure that the converter rotates in line and evenly with the flexplate. The threads on the studs or lugs should be clean and not damaged. They should also be tightly seated into the torque converter. Also check the shoulder area around the lugs and studs for cracked welds or other damage. If any damage is found, the converter should be replaced. An exception to this is when the internal threads of a drive lug are damaged. These can often be repaired by tapping the threads or by installing a threaded insert. Also inspect the converter attaching bolts or nuts, and replace them if they are damaged.

Inspect the flexplate for signs of damage, warpage, and cracks. Check the condition of the teeth on the starter ring gear. Replace the flexplate if there is any evidence of damage. Check the hub of the torque converter. It should be smooth, with no signs of wear. If the hub is worn, carefully inspect the oil pump drive and replace the torque converter. Light scratches, burrs, nicks, or scoring marks on the hub surface can be polished with a fine crocus cloth. Be careful not to allow dirt to enter into the converter while polishing the hub. If the hub has deep scratches or other major imperfections, the converter should be replaced.

In general, a torque converter should also be replaced if it has fluid leakage from its seams or welds, loose drive studs, worn drive stud shoulders, stripped drive stud threads, a heavily gouged hub, or excessive runout.

Transaxles that do not have their oil pump driven directly by the torque converter use a driveshaft that fits into a support bushing inside the converter's hub. This bushing should be checked for wear. To do this, measure the inside diameter of the bushing and the outside diameter of the driveshaft. The difference between the two measurements is the amount of clearance. If the clearance is excessive, the bushing should be replaced.

2. Gear Train, Shafts, Bushings, Oil Pump, and Case (4 Questions)

Task 1 Inspect, measure, and replace oil pump housing and parts.

Inspect the pump bore for scoring on both its bottom and sides. A converter that has had a tight fit at the pilot hub could hold the converter drive hub and inner gear too far into the pump, causing cover scoring. A front pump bushing that has too much clearance may allow the gears to run off center, causing them to wear into the crescent and/or the sides of the pump body.

The stator shaft should be inspected for looseness in the pump cover. This can be done while you check for interference inside a torque converter. The shaft's splines and bushing should also be carefully looked at. If the splines are distorted, the shaft and the pump cover should be replaced. The bushings control oil flow through the converter and cooler, and their fit must be checked. Bushings must be tight inside the shaft and provide the input shaft with a clearance.

Inspect the gears and pump parts for deep nicks, burrs, or scratches. Examine the pump housing for abnormal wear patterns. The fit of each gear into the pump body, as well as the centering effect of the front bushing, controls oil pressure loss from the high pressure side of the pump to the low pressure input side. Scoring or body wear will greatly reduce this sealing capability.

On positive displacement pumps, use a feeler gauge to measure the clearance between the outer gear and the pump pocket in the pump housing. Also, check the clearance between the outer pump gear teeth and the crescent, and between the inner gear teeth and the crescent. Compare these measurements to the specifications. Use a straightedge and feeler gauge to check gear side clearance and compare the clearance to the specifications. If the clearance is excessive, replace the pump.

Variable displacement vane type pumps require different measuring procedures. However, the inner pump rotor to converter drive hub fit is checked in the same way as described for the other pumps. The pump rotor, vanes, and slides are originally selected for size during assembly at the factory. Changing any of these parts during overhaul can destroy this sizing and possibly the body of the pump. You must maintain the original sizing and possibly the body of the pump. You must maintain the original sizing if any parts are found to need replacement. These parts are available in select sizes for just this reason.

The vanes are subject to wear, as well as cracking and subsequent breakage. The outer edge of the vanes should be rounded, with no flattening. These pumps have an aluminum body and cover halves; any scoring indicates that they should be replaced.

Inspect the reaction shaft's seal rings. If the rings are made of cast iron, check them for nicks, burrs, or uneven patterns, and replace them if they are damaged. Make sure the rings are able to rotate in their grooves. Check the clearance between the reaction shaft support ring groove and the seal ring. If the seal rings are the Teflon™ full circle type, cut them out and use the required tools to replace them.

The outer area of most pumps utilizes a rubber seal. Check the fit of the new seal by making sure the seal sticks out a bit from the groove in the pump. If it does not, it will leak. The seal at the front of the pump is always replaced during overhaul. Most of these seals are the metal clad lip seal type. Care must be taken to avoid damage to the seating area when removing the old seal.

Check the area behind the seal to be sure the drainback hole is open to the sump. If this hole is clogged, the new seal will possibly blow out. The drainback hole relieves pressure behind the seal. A loose fitting converter drive hub bushing can also cause the front pump seal to blow out.

Task 2 Check end play and/or preload; determine needed service.

While a transmission is in operation, the gears, shafts, and bearings are subjected to loads and vibrations. Because of this, the drive train must normally be adjusted for the proper fit between parts. These adjustments require the use of precision measuring tools. There are three basic adjustments that are made when reassembling a unit or when a problem suggests that readjustment is necessary. Adjusting the clearance or play between two gears in mesh is referred to as adjusting the backlash. End play adjustments limit the amount of end-to-end movement of a shaft. Preload is an adjustment made to put a load on an assembly in order to offset the load the assembly will face during operation.

Backlash is the clearance between two gears in mesh. Excessive backlash can be caused by worn gear teeth, the improper meshing of teeth, or bearings that do not support the gears properly. Excessive backlash can result in severe impact on the gear teeth from sudden stops or directional changes of the gears, which can cause broken gear teeth and gears. Insufficient backlash causes excessive overload wear on the gear teeth and could cause premature gear failure. Backlash is measured with a dial indicator mounted so that its stem is in line with the rotation of the gear and perpendicular to the angle of the teeth. One gear is then moved in both directions while the other gear is held. The amount of movement on the dial indicator equals the amount of backlash present. The proper placement of shims on a gear shaft is the normal procedure for making backlash adjustments.

End play refers to the measurable axial or end-to-end looseness of a bearing. End play is always measured in an unloaded condition. To check end play, a dial indicator is mounted against the outer side gear or the end of a shaft. The gear or shaft is then pried in both directions and the readings noted. The difference between the two readings is the amount of end play. Shims or adjusting nuts are used to adjust end play.

When normal operating loads are great, gear trains are often preloaded to reduce the deflection of parts. The amount of preload is specified in service manuals and must be corrected for the design of the bearings and the strength of the parts. If bearings are excessively preloaded, they will heat up and fail. When bearings are set too loose, the shaft will wear rapidly due to the great amounts of deflection it will experience. Gear trains are preloaded by shims, thrust washers, or adjusting nuts, or by using double race bearings. Preload adjustments are normally checked by measuring turning effort with a torque wrench.

Task 3 Inspect, measure, and replace thrust washers and bearings.

The purpose of a thrust washer is to support a thrust load and keep parts from rubbing together, thus preventing premature wear on parts, such as planetary gearsets. Selective thrust washers come in various thicknesses to take up clearances and adjust shaft end play.

Flat washers and bearings should be inspected for scoring, flaking, and wear through to the base material. Flat washers should also be checked for broken or worn tabs. These tabs are critical for holding the washer in place. On metal flat thrust washers, the tabs may appear cracked at the bend of the tab. This is a normal appearance due to the characteristics of the material used to manufacture them. Plastic thrust washers will not show wear unless they are damaged. The only way to check their wear is to measure the thickness with a micrometer and compare it to a new part. All damaged and worn thrust washers and bearings should be replaced.

Use a petroleum jelly type lubricant to hold thrust washers in place during assembly. This will keep them from falling out of place, which will affect end play. Besides petroleum jelly, there are special greases designed just for automatic transmission assembly that will work fine.

All bearings should be checked for roughness before and after cleaning. Carefully examine the inner and outer races, and the rollers, needles, or balls for cracks, pitting, etching, or signs of overheating.

Task 4 Inspect and replace shafts.

Carefully examine the areas on all shafts that ride in a bushing, bearing, or seal. Also inspect the splines for wear, cracks, or other damage. A quick way to determine spline wear is to fit the mating splines and check for lateral movement.

Shafts are checked for scoring in the areas where they ride in bushings. Since the shaft is much harder than the bushings, any scoring on the shaft indicates a lack of lubrication at that point. The affected bushing should appear worn into the backing metal. Because shaft-to-bearing fit is critical to correct oil travel throughout the transmission, a scored shaft should be replaced. Lubricating oil is carried through most shafts, and an internal inspection for debris is necessary. A blocked oil delivery hole can starve a bushing, resulting in a scored shaft. The internal oil passage of a shaft may not be able to be visually inspected, and only observation during cleaning will give an indication of the openness of the passage. Washing the shaft passage out with a solvent and possibly running a piece of small diameter wire through the passage will dislodge most particles. Be sure to check that the ball closes off the end of the shaft, if the shaft is so equipped, and is securely in place. A missing ball could be the cause of burned planetary gears and scored shafts due to a loss of oil pressure. Any shaft that has an internal bushing should be inspected as described earlier. Replace all defective parts as necessary.

Input and output shafts can be solid, drilled, or tubular. The solid and drilled shafts are supported by bushings, so the bushing journals of the shaft should be free of noticeable wear at these points. Small scratches can be removed with 300-grit emery paper. Grooved or scored shafts require replacement. The splines should not show any sign of waviness along their length. Check drilled shafts to be sure the drilled portion is free of any foreign material. Wash out the shaft with solvent and run a small diameter wire through the shaft to dislodge any particles. After running the wire through the opening, wash out the shaft once more and blow it out with compressed air.

If the shaft has a check ball, be certain the ball seats in the correct direction. Some shafts have a ball pressed into one end to block off one end of the shaft. This is used to hold oil in the shaft so the oil is diverted through holes in the side of the shaft. These holes supply oil to bushings, one-way clutches, and planetary gear. If the ball does not fully block the end of the shaft, oil pressure can be lost, causing failure of these components. Some shafts may be used to support another shaft. The output shaft uses the rear of the input shaft to center and support itself. The small bushing found in the front end of the output shaft should always be replaced on these transmissions during rebuilding.

Task 5 Inspect oil delivery seal rings, ring grooves, sealing surface areas, feed pipes, orifices, and encapsulated check valves (balls).

At times, leaks may be from sources other than seals. Leakage could be from a worn gasket, loose bolts, cracked housing, or loose line connections. Inspect the outside sealing area of the seal to see if it is wet or dry. If it is wet, see whether the oil is running out or if it is merely a lubricating film. Check both the inner and outer parts of the seals for wet oil.

While removing a seal, inspect the sealing surface, or lips, before cleaning it. Look for signs of unusual wear, warpage, cuts and gouges, or particles embedded in the seal. On spring loaded lip seals, make sure that the spring is seated around the lip and that the lip was not damaged when first installed. If the seal's lip is hardened, this was probably caused by heat from either the shaft or the fluid.

If the seal is damaged, check all shafts for roughness, especially at the seal contact area. Look for deep scratches or nicks that could have damaged the seal. Determine if shaft spline, keyway, or a burred end could have caused a nick or cut in the seal lip during installation. Inspect the bore into which the seal was fitted. Look for nicks and gouges that could create a path of oil leakage. A coarsely machined bore can allow oil to seep out through a spiral path. Sharp corners at the bore edges can score the metal case of the seal when it is installed. These scores can make a path for oil leakage.

Task 6 Inspect and replace bushings.

Bushings should be inspected for pitting and scoring. Always check the depth to which bushings are installed and the direction of the oil groove, if so equipped, before you remove them. Many bushings that are used in the planetary gearing and output shaft area have oiling holes in them. Be sure to line these up correctly during installation or you may block off oil delivery and destroy the gear train. If any damage is evident on the bushing, it should be replaced.

Bushing wear can be checked directly, as well as checked by observing the lateral movement of the shaft that fits into the bushing. Any noticeable lateral movement indicates wear, and the bushing should be replaced. The amount of clearance between the shaft and the bushing can be checked with a wire type feeler gauge. Insert the wire between the shaft and the bushing. If the gap is greater than the maximum allowable gap, the bushing should be replaced. You can check this fit by measuring the inside diameter of the bushing and the outside diameter of the shaft with a vernier caliper or micrometer. This is a critical fit throughout the transmission and especially at the converter drive hub.

Most bushings are press fit into a bore. To remove them, they are driven out of the bore with a properly sized bushing tool. Some bushings can be removed with a slide hammer fitted with an expanded or threaded fixture that grips to the inside of the bushing and collapses it. Once collapsed, the bushing can easily be removed with a pair of pliers. Small bore bushings that are located in areas where it is difficult to use a bushing tool can be removed by tapping the inside bore of the bushing with threads that match a selected bolt, which fits into the bushing. After the bushing has been tapped, insert the bolt and use a slide hammer to pull the bolt and bushing out of its bore.

Whenever possible, all new bushings should be installed with the proper bushing driver. The use of these tools prevents damage to the bushing and allows for proper seating of the bushing into its bore.

Task 7 Inspect and measure planetary gear assembly; replace parts as necessary.

A close inspection of the planetary gearset is a must to eliminate the possibility of causing noises in a newly rebuilt unit. All planetary gear teeth should be inspected for chips or stripped teeth. Any gear that is mounted to a splined shaft must have its splines checked for mutilation or shifted splines. The planetary gears used in automatic transmissions are helical type gears, like the ones used in most manual transmissions. This type of gear provides low noise in operation, but makes it necessary to check the end play of individual gears during inspection. The helical cut makes the gears thrust to one side during inspection. This can put a lot of load on the thrust washers and may wear them beyond specification.

Look first for obvious problems like blackened gears or pinion shafts. These conditions indicate severe overloading and require that the carrier be replaced. Occasionally, the pinion gear and shaft assembly can be replaced individually. When looking at the gears themselves, note that a bluish color can be a normal condition, as this is part of a heat-treating process used during manufacturing. Check the planetary pinion gears for loose bearings. Check each gear individually by rolling it on its shaft to feel for roughness or binding of the needle bearings. Wiggle the gear to be sure it is not loose on the shaft and to feel for roughness or binding of the needle bearings. Looseness will cause the gear to whine when it is loaded. Also, inspect the gears and teeth for chips or imperfections, as these will also cause whine.

Check the gear teeth around the inside of the front planetary ring gear. Check the fit between the front planetary carrier to the output shaft splines. Remove the snapring gear. Check the fit between the front planetary ring gear. Examine the thrust washer and the outer splines of the front drum for burrs and distortion. The rear clutch friction discs must be able to slide on these splines during engagement and disengagement. With the snapring removed, the front planetary carrier can be removed from the ring gear. Check the planetary carrier gears for end play by placing a feeler gauge between the planetary carrier and the planetary pinion gear. Compare the end play to specifications. On some Ravigneaux units, the clearance at both ends of the long pinion gears must also be checked and compared to specifications.

Check the splines of the sungear. Sungears should have their inner bushings inspected for looseness on their respective shafts. Also check the fit of the sun shell to the sungear. The shell can crack where the gears mate with the shell. The sun shell should also be checked for a bell mouthed condition where it is tabbed to the clutch drum. Any shell with a variation from true round should be considered junk and should not be used. Look at the tabs and check for the best fit into the clutch drum slots. This involves trial fitting the shell and drum at all possible combinations and marking the point where they fit the tightest. A snug fit here will eliminate bell mouthing due to excess play at the tabs. It can also reduce engagement noise in reverse, second, and fourth gears. This excess play allows the sun shell tabs to strike the clutch drum tabs as the transmission shifts from first to second or when the transmission is shifted into reverse.

The gear carrier should have no cracks or other defects. Replace any abnormal or worn parts. Check the thrust washer for excessive wear and, if required, correct the input shaft thrust clearance by using a washer with the correct thickness. Determine the correct thickness by measuring the thickness of the existing thrust washer and comparing it to the measured end play. Now move the gear back and forth to check its end play. Some shop manuals will give a range for this check, but if none can be found, you can figure about 0.007 to 0.025 inch (0.178 to 0.635 mm) as an average amount. All the pinions should have about the same end play.

Task 8 Inspect, repair, and replace case(s), bores, passages, bushings, vents, mating surfaces, and dowel pins.

The transmission case should be thoroughly cleaned and all passages blown out. After the case has been cleaned, all the bushings, fluid passages, bolt threads, clutch plate splines, and the governor bore should be checked. The passages can be checked for restrictions and leaks by applying compressed air to each one. If the air comes out the outer end, there is no restriction. To check for leaks, plug off one end of the passage and apply air to the other. If pressure builds in that passage, there are probably no leaks in it.

Modern transmission cases are made of aluminum, primarily to save weight. Aluminum is a soft material that can be deformed, scratched, cracked, or scored much more easily than cast iron. Special attention should be given to the following areas: the clutch, the oil pump, the servo, and the accumulator bores. All bores should be smooth to avoid scratching or tearing the seals. The servo piston could also hang up in a bore that is deeply scored. Check the fit of the servo piston in the bore without the seal, if possible, to be sure it has free travel. There should be no tight spots or binding over the whole range of travel. Any deep scratches or gouges that cause binding of the piston will require case replacement.

Case-mounted accumulator bores are checked the same as servo bores. The oil pump bore at the front of the case should be free of any scratches that would keep the O-ring from sealing the outer diameter of the pump to the front of the case. Case-mounted hydraulic clutch bores are prone to the same problems as servo bores. Look for any scratches or gouges in the sealing area that would affect the rubber seals. It is possible to damage these areas during disassembly, so be careful with tools used during overhaul.

Sealing surfaces of the cue should be inspected for surface roughness, nicks, or scratches where the seals ride. Any problems found in servo bores, clutch drum bores, or governor support bores can cause pressure leakage in the affected circuit. Imperfections in steel or cast-iron parts can usually be polished out with a crocus cloth. Care should be taken so as not to disturb the original shape of the bore. Under no circumstances should sandpaper be used. Sandpaper will leave too deep a scratch in the surface. Use a crocus cloth inside clutch drums to remove the polished marks left by the cast-iron sealing rings. This will help the new rings rotate with the drum as designed. As a rule, all sealing rings, either cast-iron or Teflon™, are replaced during overhaul, as this gives the desired sealing surface required for proper operation.

Passages in the case guide the flow of fluid through the case. Although not common, porosity in this area can cause cross-tracking of one circuit to another. This can cause bind-up (two gears at once) or a slow bleed of pressure in the affected circuit, which can lead to slow burnout of a clutch or band. If this is suspected, try filling the circuit with solvent and watching to see if the solvent disappears or leaks away. If the solvent goes down, you will have to check each part of the circuit to find where the leak is. Be sure to check that all necessary check balls were in position during disassembly.

Small screens found during teardown should be inspected for foreign material. These screens are used to prevent valve hang-up at the pressure regulator and governor. Most screens can be removed easily. Care should be taken when cleaning because some solvents will destroy the plastic screens. Low air pressure can be used to blow the screens out in a reverse direction.

Bushings in a transmission case are normally found in the rear of the case and require the same inspection and replacement techniques as other bushings in the transmission. Always be sure that the oil passages to a pressure fed bushing or bearing are open and free of dirt and foreign material. It does no good to replace a bushing without checking to be sure it has good oil flow.

Vents are located in the pump body or transmission case and provide for equalization of pressure in the transmission. These vents can be checked by blowing low pressure air through them, squirting solvent brake cleaning spray through them, or by pushing a small diameter wire through the vent passage. A clean, open passage is all you need to verify proper operation.

Task 9 Inspect, repair, and replace transaxle drive chains, sprockets, gear bearings, and bushings.

The drive chains used in some transaxles should be inspected for side play and stretch. These checks are made during disassembly and should be repeated as a double check during reassembly. Chain deflection is measured between the centers of the two sprockets. Typically, very little deflection is allowed.

Deflect the chain inward on one side until it is tight. Mark the housing at the point of maximum deflection. Then deflect the chain outward on the same side until it is tight. Again mark the housing in line with the outer edge of the chain at the point of maximum deflection. Measure the distance between the two marks. If this distance exceeds specifications, replace the drive chain.

Be sure to check for an identification mark on the chain during disassembly. These can be painted or have dark colored links, which may indicate either the top or the bottom of the chain, so be sure you remember which side was up.

The sprockets should be inspected for tooth wear at the point where they ride. If the chain was found to be slack, it may have worn the sprockets in the same manner as the engine timing gears wear when the timing chain stretches. A slightly polished appearance on the face of the gears is normal.

The bearings and bushings used on the sprockets need to be checked for damage. The radial needle thrust bearings must be checked for any deterioration of the needles and cage. The running surface in the sprocket must also be checked, as the needles may pound into the gear's surface during operation. The bushings should be checked for any signs of scoring, flaking, or wear. Replace any defective parts. The removal and installation of the chain drive assembly of some transaxles requires that the sprockets be spread slightly apart. The key to doing this is to spread the sprockets just the correct amount. If they are spread too far, they will not be easy to install or remove.

Task 10 Inspect, measure, repair, and adjust or replace transaxle final drive components.

Transaxle final drive units should be carefully inspected. Examine each gear, thrust washer, and shaft for signs of damage. If the gears are chipped or broken, they should be replaced. Also inspect the gears for signs of overheating or scoring on the bearing surface of the gears.

Final drive units may be helical gear or planetary gear units. The helical type should be checked for worn or chipped teeth, overloaded tapered roller bearings, and excessive differential side gear and spider gear wear. Excessive play in the differential is a cause of engagement clunk. Be sure to measure the clearance between the side gears and the differential case, and to check the fit of the spider gears on the spider gear shaft. Proper clearances can be found in the appropriate shop manual. It is possible that the side bearings of some final drive units are preloaded with shims. Select the correct size shim to bring the unit into specifications. With a torque wrench, measure the amount of rotating torque. Compare your readings against specifications.

If the bearing preload and end play are fine and the bearings are in good condition, the parts can be reused. However, always install new seals during assembly. It should be noted that these bearings function the same as RWD rear axle side bearings, and, should not set the preload to the specifications for a new bearing. Used bearings should be set to the amount found during teardown or about one-half the preload of a new bearing.

Planetary-type final drives are also checked for the same differential case problems that the helical type would encounter. The planetary pinion gears need to be checked for looseness or roughness on their shafts and for end play. Any problems found normally result in the replacement of the carrier as a unit since most pinion bearings and shafts are sold as separate parts. Again, specifications for these parts are found in the shop manual.

Planetary type final drives, like helical final drives, are available in more than one possible ratio for a given type of transaxle, so care should be taken to assure that the same gear ratios are used during assembly. This is not normally a problem when overhauling a single unit; however, in a shop where many transmissions are being repaired, it is possible to mix up parts, causing problems during the rebuild.

3. Friction and Reaction Units (4 Questions)

Task 1 Inspect clutch assembly; replace as necessary.

Once a clutch assembly has been taken apart, you may wish to inspect the clutch components or continue to disassemble the remainder of the clutch units in the transmission. If you choose the latter, make sure you keep the parts of each clutch separate from the others.

Clean the components of the clutch assembly. Make sure all clutch parts are free of any residue of varnish, burned disc facing material, or steel filings. Take special care to wash out any foreign material from the inside of the drums and hub disc splines. If left in, the material can be washed out by the fresh transmission fluid and sent through the transmission. This can ruin the rebuild.

The clutch splines must be in good shape with no excessively rounded corners or shifted splines. Test their fit by trial-fitting three new clutch discs on the splines. Move the discs up and down the splines to check for binding. If they bind, this can cause dragging of the discs during a time when they should be free-floating. Replace the hubs if the discs drag during this check. Check the spring retainer; it should be free-floating. Replace the hubs if the discs drag during this check. Check the spring retainer; it should be flat and not distorted at its inner circumference. Check all springs for height, cracks, and straightness. Any springs that are not the correct height or that are distorted should be replaced. Many retainers have springs attached to them by crimping. This speeds up production at the assembly line. Turning this type of retainer upside down is a quick check of spring length. Closely examine the Belleville spring for signs of overheating or cracking, and replace it if it is damaged.

The steel plates should be checked to be sure they are flat and not worn too thin. Check all steel plates against the thickest one in the pack or a new one. Most steel plates will have an identification notch or mark on the outer tabs. If the plates pass inspection, remove the polished surface finish so the steel plates are ready for reuse. The steel plates should also be checked for flatness by placing one plate on top of the other and checking the shape on the inside and outside diameters. Clutch plates must not be warped or cone shaped. Also, check the steel plates for burning and scoring, and for damaged driving lugs. Check the grooves inside the clutch drum and check the fit of the steel plates, which should travel freely in the grooves.

Close inspection of the friction discs is simple. The disc will show the same type of wear as bands will. Disc facing should be free of chunking, flaking, and burnt or blackened surfaces. Discs that are stripped of their facing have been overheated and subject to abuse. In some cases, the friction discs and steel plates can be welded together. This occurs when the facing comes off the disc due to a loosening of the facing's bonding because of extreme heat. As the facing comes off, metal-to-metal contact is made and the disc and plate fuse together. This may lock the clutch in an engaged condition. Depending on which clutch is affected, drivability problems can include driving in neutral, binding up in reverse, starting in direct drive, binding up in second, and other problems that are not that common.

If the discs do not show any signs of deterioration, squeeze each disc to see if fluid is still trapped in the facing material. If fluid comes to the surface, the disc is not glazed. Glazing seals off the surface of the disc and prevents it from holding fluid. Holding fluid is basic to proper disc operation. It allows the disc to survive engagement heat, which would otherwise burn the facing and cause glazing. Fluid stored in the friction material cools and lubricates the facing as it transfers heat to the steel plate and also carries heat away as some oil is spun out of the clutch pack by centrifugal force. This helps avoid the scorching and burning of the disc. The clutch disc must not be charred, glazed, or heavily pitted. If a disc shows signs of flaking or if friction material can be scraped off easily, replace the disc. A black line around the center of the friction surface also indicates that the disc should be replaced. Examine the teeth on the inside of each friction disc for wear and other damage.

Wave plates are used in some clutch assemblies to cushion the application of the clutch. These should be inspected for cracks and other damage. Never mix wave plates from one clutch assembly with another. As an aid in assembly, most wave plates will have different identifying marks.

Task 2 Measure and adjust clutch pack clearance.

The clearance check of a clutch pack is critical for correct transmission operation. Excessive clearance causes delayed gear engagements while too little clearance causes the clutch to drag. Adjusting the clearance of multiple-disc clutches can be done with the large outer snapring in place.

With the clutch pack and pressure plate installed, use a feeler gauge to check the distance between the pressure plate and the outer snapring. Clearances can also be measured between the backing plate and the uppermost friction disc. If the clutch pack has a waved snapring, place the feeler gauge between the flat pressure plate and the wave of the snapring farthest away from the pressure plate. Compare the distance to specifications. Attempt to set the pack clearance to the smallest dimension shown in the chart.

Clearance can also be checked with a dial indicator and hook tool. The hook tool is used to raise one disc from its downward position, and the amount that it is able to move is recorded on the dial indicator. This represents the clearance.

If the clearance is greater than specified, install a thicker snapring to take up the clearance. If the clutch clearance is insufficient, install a thinner snapring.

Task 3 Air test the operation of clutch and servo assemblies.

After the clearance of the clutch pack is set, perform an air test on each clutch. This test will verify that all of the seals and check balls in the hydraulic components are able to hold and release pressure.

Air checks can also be made with the transmission assembled. This is the absolute best way to check the condition of the circuit because there are very few components missing from the circuit. The manufacturers of different transmissions have designed test plates that are available to test different hydraulic circuits. Testing with the transmission assembled also allows for testing of the servos.

To test a clutch assembly, install the oil pump assembly with its reaction shaft support over the input shaft and slide it in place on the front clutch drum. When the clutch drums are mounted on the oil pump, all components in the circuit can be checked. If the clutch cannot be checked in this manner, blocking off apply ports with your finger and applying air pressure through the other clutch apply port will work.

Invert the entire assembly and place it in an open vise or transmission support tool. Then air test the circuit using the test hole designated for that clutch. Be sure to use low-pressure compressed air to avoid damage to the seals. High-pressure air may blow the rubber seals out of the bore or roll them on the piston.

While applying air pressure, you may notice some air escaping at the metal of the Teflon™ seal areas. This is normal, as these seals have a controlled amount of leakage designed into them. There should be no air escaping from the piston seals. The clutch should apply with a dull but positive thud. It should release quickly without any delay or binding. Examine the check ball seat for evidence of air leakage.

Task 4 Inspect one-way clutch assemblies; replace parts as necessary.

Because they are purely mechanical in nature, one-way clutches are relatively simple to inspect and test. The durability of these clutches relies on constant fluid flow during operation. If a one-way clutch has failed, a thorough inspection of the hydraulic feed circuit to the clutch must be made to determine if the failure was due to fluid starvation. The rollers and sprags ride on an overrunning state, and any loss of fluid can cause a rapid failure of the components. Sprags, by design, produce the fluid wave effect as they slide across the inner and outer races, making them somewhat less prone to damage. Rollers, due to their spinning action, tend to throw off fluid, which allows more chance for damage during fluid starvation. During the check of the hydraulic circuit, take a look at the feed holes in the races of the clutch. Use a small diameter wire and spray carburetor cleaner or brake cleaner to be certain the feed holes are clear. Push the wire through the feed holes and spray the cleaner into them. Blowing through them with compressed air after cleaning is recommended.

Roller clutches should be disassembled to inspect the individual pieces. The surface of the rollers should have a smooth finish with no evidence of any flatness. Likewise, the race should be smooth and show no sign of brinnelling, as this indicates severe impact loading. This condition may also cause the roller clutch to buzz as it overruns.

All rollers and races that show any type of damage or surface irregularities should be replaced. Check the folded springs for cracks, broken ends, or flattening out. All of the springs from a clutch assembly should have approximately the same shape. Replace all distorted or otherwise damaged springs. The cam surface, like the smooth race, must be free of all irregularities.

Sprag clutches cannot easily be disassembled, so a complete and thorough inspection of the assembly is necessary. Pay particular attention to the faces of the sprags. If the faces are damaged, the clutch unit should be replaced. Sprags and races with scored or torn faces are an indication of dry running and require the replacement of the complete unit.

Once the one-way clutch is ready for installation, verify that it overruns in the proper direction. In some cases, it is possible to install the clutch backwards, which would cause it to overrun and lock in the wrong direction. This would result in some definite drivability problems. To make sure you have installed the clutch in the correct direction, determine the direction of lockup before installing the clutch.

Task 5 Inspect and replace bands and drums.

Servicing of bands and their components includes inspection of the bands, as well as the drums that the bands wrap around. Before the introduction of overdrive automatic transmissions, most bands operated in free condition during most driving conditions. This means the band was not applied in the cruising gear range. However, many overdrive automatic transmissions use a band in the overdrive cruise range, which puts an additional load on the band and subsequently causes additional wear on the band. For this reason, a thorough inspection of the bands is very important.

The bands in a transmission will be either single or double wrap, depending on the application. Both types can be the heavy-duty cast-iron type or the normal strap type. The friction material used on clutches and bands is quite absorbent. This characteristic can be used to tell if there is much life left in the lining. Simply squeeze the lining with your fingers to see if any fluid appears. If fluid appears, this tells you the lining can still hold fluid and has some life left in it. It is hard to tell exactly how long the band will last, but at least you have an indication that it is still useable. Strap or flex type bands should never be twisted or flattened out. This may crack the lining and lead to flaking of the lining.

Band failure found during an overhaul is easy to spot. Look for chipping, cracks, burn marks, glazing, and nonuniform wear patterns and flaking. If any of these defects are apparent, the band should be replaced.

Also inspect brake band friction material for wear. If the linings show wear, carefully check the band struts, levers, and anchors for wear. Replace any worn or damaged parts. Look at the linings of heavy-duty bands to see if the lining is worn evenly. A twisted band will show taper wear on the lining. If the friction material is blackened, this is caused by an excessive buildup of heat. High heat may weaken the bonding of the lining and allow the lining to come loose from the metal portion of the band.

The drum surface should be checked for discoloration, scoring, glazing, and distortion. The drums will be either iron castings or steel stampings. Cast-iron drums that are not scored can generally be restored to service by sanding the running surface with 180-grit emery paper in the drum's normal direction of rotation. A polished surface is not desirable on cast-iron drums.

The surface of the drum must also be flat. This is not usually a problem with a cast-iron drum, but it can affect the stamped steel type drum. It is possible for the outer surface of the drum to dish outward during its normal service life. Check the drum for flatness across the outer surface where the band runs. Any dishing here will cause the band to distort as it attempts to get a full grip on the drum. Distortion of the band weakens the bond of the friction material to the band and will cause early failure due to flaking of the friction lining. A dished stamped steel drum should be replaced. Check the service manual for maximum allowable tolerances.

Sample Test for Practice

Sample Test

Please note the letter and number in parentheses following each question. They match the overview in section 4 that discusses the relevant subject matter. You may want to refer to the overview using this cross-referencing key to help with questions posing problems for you.

1. A vehicle is brought in with an automatic transmission that won't downshift from cruising speed (forced downshift). Technician A says that the throttle valve (TV) cable may be out of adjustment. Technician B says that the kickdown switch may be open. Who is right?
   A. A only
   B. B only
   C. Both A and B
   D. Neither A nor B (B.2)
2. All of the following statements about bushing inspection and measurement are true EXCEPT:
   A. a scored shaft in the bushing contact area may indicate a lack of lubrication.
   B. the shaft to bushing clearance may be measured with a wire type feeler gauge.
   C. the shaft to bushing clearance may be measured with a vernier caliper and a micrometer.
   D. normal bushing clearance is 0.015 to 0.025 inch (0.381 to 0.635 mm). (D.2.6)
3. A customer with an automatic transaxle in a vehicle has a complaint of harsh 3-4 upshifts. All the other shifts are normal. Technician A says the fourth accumulator piston may be stuck. Technician B says the pressure regulator valve is sticking. Who is right?
   A. A only
   B. B only
   C. Both A and B
   D. Neither A nor B (C.10)
4. When removing a transmission on a rear wheel drive (RWD) vehicle, remove all of the following components EXCEPT:
   A. transmission cooler lines.
   B. flexplate.
   C. torque converter inspection cover.
   D. driveshaft. (D.1.1)

5. While checking clutch discs, Technician A says the steel plates should be replaced if they are worn flat. Technician B says the friction discs should be squeezed to see if they can hold fluid. Who is right?
   A. A only
   B. B only
   C. Both A and B
   D. Neither A nor B (D.3.1)
6. All of the following problems could result in clutch disc burning EXCEPT:
   A. a sticking clutch drum check ball.
   B. a reduced clutch pack clearance.
   C. a damaged clutch piston seal.
   D. a higher-than-specified line pressure. (D.3.2)
7. A planetary gearset is being inspected. Technician A says that all splined shaft mounted gears must have the splines inspected for shifting of the splines. Technician B says that blackened gears or pinion shafts indicate overheating and should be replaced. Who is right?
   A. A only
   B. B only
   C. Both A and B
   D. Neither A nor B (D.2.7)
8. A transmission experiences repeated pump seal failure, and there are no other transmission complaints. Technician A says the pressure regulator valve may be sticking. Technician B says the drainback hole behind the seal may be plugged. Who is right?
   A. A only
   B. B only
   C. Both A and B
   D. Neither A nor B (D.2.1)
9. A computer controlled transaxle does not shift into fourth gear. Technician A says to begin diagnosis with a basic inspection of wires and hoses. Technician B says that battery voltage should be tested. Who is right?
   A. A only
   B. B only
   C. Both A and B
   D. Neither A nor B (A.2.3)
10. Technician A says that if transmission bands are squeezed and fluid appears, the band is still useable. Technician B says that the clutch drum should be inspected with the bands. Who is right?
    A. A only
    B. B only
    C. Both A and B
    D. Neither A nor B (D.3.5)

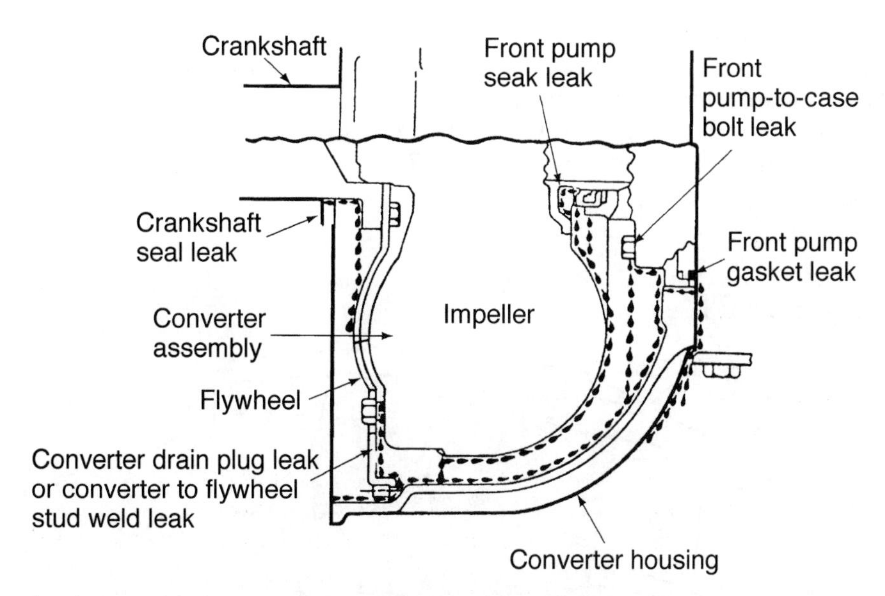

11. In the figure above, a red lubricant is leaking from the torque converter access cover. When this cover is removed, the shell of the converter is wet with fluid, but the front of the converter is dry. The cause of the problem could be:
    A. a leaking converter drain plug.
    B. a leaking rear main bearing.
    C. a loose rear main bearing.
    D. a leaking transmission oil pump seal. (A.1.3)
12. A customer complains about fluid usage in an automatic transmission, and there are no visible signs of fluid leaks. Technician A says the transmission vent may be plugged or restricted. Technician B says the vacuum modulator diaphragm may be leaking. Who is right?
    A. A only
    B. B only
    C. Both A and B
    D. Neither A nor B (A.1.1)
13. A customer complains about slippage between shifts. Technician A says a low fluid level can cause low pressures, which cause slippage between shifts. Technician B says a low fluid level can be an indicator that there is an external fluid leak. Who is right?
    A. A only
    B. B only
    C. Both A and B
    D. Neither A nor B (A.1.1)
14. During a stall test, the engine speed (RPM) is less than specified. Technician A says the turbine in the torque converter may be defective. Technician B says some of the transmission clutches may be slipping.Who is right?
    A. A only
    B. B only
    C. Both A and B
    D. Neither A nor B (A.1.5)

15. After a noncomputer-controlled transaxle and valve body overhaul, the transaxle does not complete a 1-2 upshift, and shifts from first gear to third gear. All other shifts are normal, and this problem was not present before the overhaul. Technician A says the valve body torque may be excessive. Technician B says the governor pressure may be too low. Who is right?
    A. A only
    B. B only
    C. Both A and B
    D. Neither A nor B (C.8)

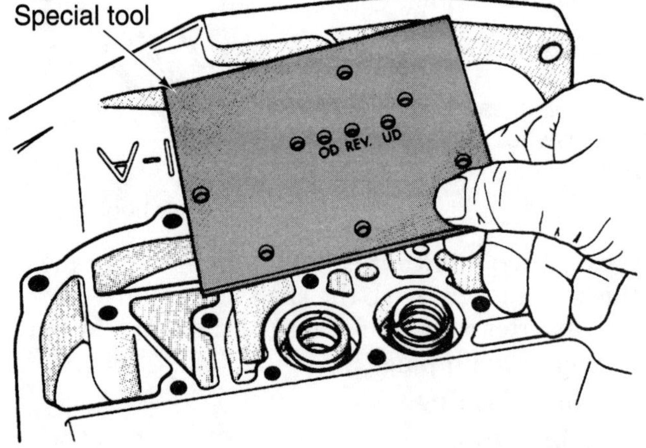

16. In the figure above, Technician A says an air test can be used to check servo action. Technician B says an air test can be used to check for external leaks. Who is right?
    A. A only
    B. B only
    C. Both A and B
    D. Neither A nor B (A.1.7)

17. While discussing various sensors used with an electronic transmission control system, Technician A says potentiometers are typically used to measure temperature changes. Technician B says vacuum modulators are used to measure engine load. Who is right?
    A A only
    B. B only
    C. Both A and B
    D. Neither A nor B (A.2.4)

18. All transmission pressures are normal at idle speed, but low at wide open throttle. Technician A says the Throttle Valve (TV) cable may need adjusting. Technician B says the vacuum modulator may be defective. Who is right?
    A. A only
    B. B only
    C. Both A and B
    D. Neither A nor B (A.1.4)

19. While discussing a pressure test, Technician A says this test is the most valuable diagnostic check for slippage in one gear. Technician B says the test can identify the cause of late or harsh shifting. Who is right?
    A. A only
    B. B only
    C. Both A and B
    D. Neither A nor B (A.2.2)

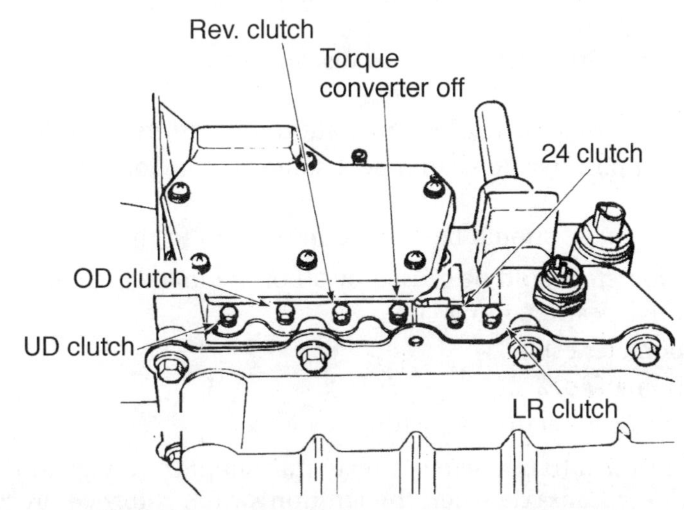

20. During an air pressure test on the reverse clutch shown in the figure above, the clutch application is not heard and a hissing noise is evident. Technician A says the transmission case may be cracked. Technician B says the reverse clutch drum may be cracked. Who is right?
    A. A only
    B. B only
    C. Both A and B
    D. Neither A nor B (A.1.7)
21. While discussing band operation, Technician A says a servo is a hydraulically operated piston assembly used to apply the band. Technician B says an accumulator is a hydraulic piston assembly that helps a servo to quickly apply the band. Who is right?
    A. A only
    B. B only
    C. Both A and B
    D. Neither A nor B (C.9, C.10)
22. If a gear selection linkage on an automatic transmission is misadjusted all of the following could happen EXCEPT:
    A. poor gear engagement.
    B. low fluid pressure.
    C. gear indicator reading incorrect.
    D. leak at converter hub. (B.1)
23. While performing a clutch plate clearance measurement, the dial indicator reading is more than specified. To correct this problem, install:
    A a thicker selective thrust washer.
    B new friction and steel clutch plates.
    C. a thicker snap ring.
    D. a new clutch drum. (D.3.2)

24. With the engine idling, the automatic transmission fluid (ATF) flow through a transmission cooler should be:
    A. one quart in 60 seconds.
    B. one pint in 60 seconds.
    C. one pint in 20 seconds.
    D. one quart in 20 seconds. (C.5)
25. All the following statements about transmission/transaxle removal and replacement are true EXCEPT:
    A. the negative battery cable should be disconnected prior to transmission removal.
    B. the driveshaft should be marked in relation to the differential flange.
    C. the engine support fixture should be installed before loosening the transaxle to engine bolts.
    D. the front drive axles should be marked in relation to the front hubs. (D.1.1)
26. An erratic speedometer could be caused by all of the following defects EXCEPT:
    A. a missing drive gear retaining clip.
    B. a dry speedometer cable.
    C. worn speedometer gears.
    D. a worn driven gear retaining bushing. (D.3.5)
27. A computer-controlled transaxle has a relay that supplies voltage to the solenoids and switches in the transaxle when the ignition switch is turned on. The computer senses a defect in the input speed sensor and does not close the relay. Under this condition, the transaxle operates in:
    A. first gear and reverse.
    B. second gear and reverse.
    C. first and second gear.
    D. second and third gear. (C.12)
28. Technician A says most vibrations problems are caused by an unbalanced torque converter. Technician B says vibration problems can be caused by a faulty output shaft. Who is right?
    A. A only
    B. B only
    C. Both A and B
    D. Neither A nor B (A.1.2)
29. Technician A says an improper shift linkage adjustment may cause premature transmission clutch failure. Technician B says an improper shift linkage adjustment may cause higher than normal fluid pressure. Who is right?
    A. A only
    B. B only
    C. Both A and B
    D. Neither A nor B (B.1)
30. While discussing a battery capacity test with the battery temperature at 70°F (21°C), Technician A says the battery is satisfactory if the voltage remains above 9.6 Volts. Technician B says the battery discharge rate is calculated by multiplying two times the battery reserve capacity rating. Who is right?
    A. A only
    B. B only
    C. Both A and B
    D. Neither A nor B (A.2.5)

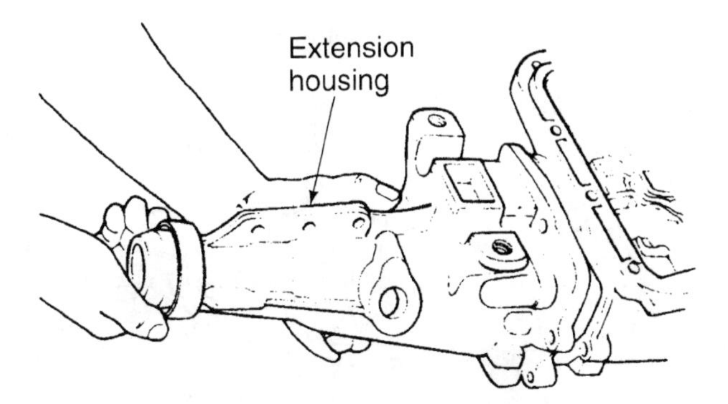

31. The extension housing shown above experiences repeated extension housing rear seal failures. Technician A says to check the transmission vent for blockage. Technician B says to replace the extension housing bushing. Who is right?
    A. A only
    B. B only
    C. Both A and B
    D. Neither A nor B (C.4)
32. The following seals are common sources for leaks in an auto transmission EXCEPT:
    A oil pan seal.
    B. speedometer drive seal.
    C. electrical components installed into transmission case.
    D. accumulator seal. (C.3)
33. All of the following statements about a parking pawl are true EXCEPT:
    A. the parking pawl locks the input shaft.
    B. the parking pawl is mechanically operated.
    C. the parking pawl's projection is engaged in a notched drum.
    D. the parking pawl is retained on a pivot pin. (C.11)
34. A transmission is diagnosed to have a sticking valve in the valve body. Technician A says that poor maintenance and overheating will cause sticking valves. Technician B says that the valve body must be removed to address this problem. Who is right?
    A. A only
    B. B only
    C. Both A and B
    D. Neither A nor B (C.7)
35. The shifts in an automatic transaxle occur at a higher speed than specified. All of the following items could be the cause of the problem EXCEPT:
    A. a sticking governor valve.
    B. excessive governor spring tension.
    C. worn governor weights and pins.
    D. weak governor spring tension. (C.2)
36. The clearance between the valves and matching valve body bores should not exceed:
    A. 0.001 inch (0.025 mm).
    B. 0.003 inch (0.084 mm).
    C. 0.005 inch (0.127 mm).
    D. 0.008 inch (0.203 mm). (C.7)

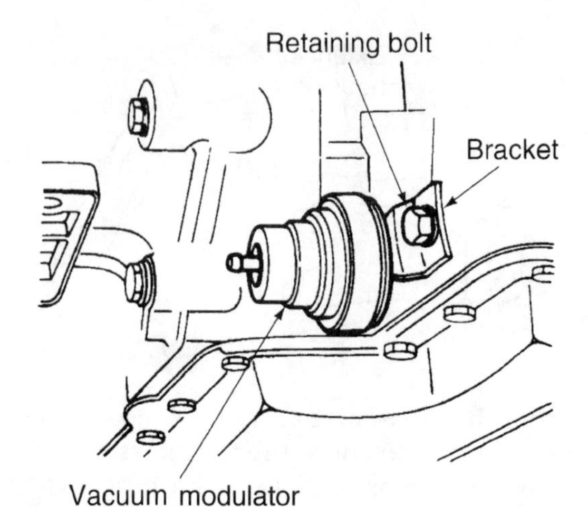

37. As shown above, a vacuum modulator is being inspected on an automatic transmission. Technician A says that the vacuum modulator should be receiving venturi vacuum. Technician B says that if fluid leaks from the modulator with the vacuum line disconnected, the modulator should be replaced. Who is right?
    A. A only
    B. B only
    C. Both A and B
    D. Neither A nor B (C.1)
38. While discussing adjustments on an automatic transmission, Technician A says that the three basic adjustments are backlash, end play, and preload. Technician B. says that excessive backlash can cause broken gear teeth. Who is right ?
    A. A only
    B. B only
    C. Both A and B
    D. Neither A nor B (D.2.2)
39. An engine has a clunking noise that usually occurs during deceleration and sometimes with the engine idling. The cause of this noise could be loose:
    A. main bearings.
    B. converter to flexplate bolts.
    C. connecting rod bearings.
    D. piston pins. (D.1.4)
40. While checking the condition of a car's automatic transmission fluid (ATF), Technician A says if the fluid has a dark brownish or blackish color and/or a burned odor, the fluid has been overheated. Technician B says if the fluid has a milky color, this indicates that engine coolant has been leaking into the transmission's cooler. Who is right?
    A. A only
    B. B only
    C. Both A and B
    D. Neither A nor B (C.5)

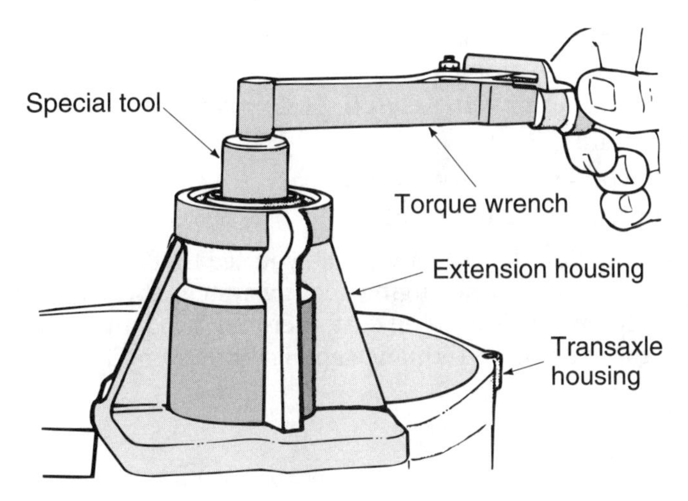

41. In the figure above, the Technician is most likely measuring:
    A. end play.
    B. differential turning torque.
    C. side spider gear wear.
    D. engagement clunk. (D.2.10)

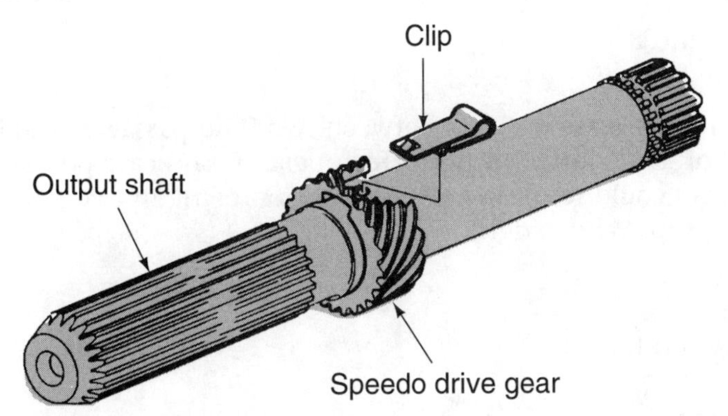

42. In the figure above, speedometer gear problems can be caused by all of the following EXCEPT a:
    A. faulty cable.
    B. faulty drive/driven gear.
    C. faulty speedometer.
    D. faulty tailshaft. (C.6)

43. Technician A says a blocked oil delivery passage will cause the shaft to score. Technician B says if the shaft is fitted with a ball and the ball does not seat properly, loss of oil pressure will result. Who is right?
    A. A only
    B. B only
    C. Both A and B
    D. Neither A nor B (D.2.4)

44. An improper band adjustment may cause:
    A. shifts at a lower vehicle speed than specified.
    B. transmission slipping in some gears.
    C. shifts at a higher vehicle speed than specified.
    D. transmission slipping in all gears. (B.3)

45. A torque converter clutch does not lock up at any vehicle speed or engine temperature. The cause of this problem could be:
    A. a defective torque converter clutch (TCC) solenoid.
    B. a defective stator clutch.
    C. a leaking second speed servo piston seal.
    D. a sticking pressure regulator valve. (A.1.6)

46. After checking the fluid level of a vehicle, you see that the fluid looks and smells burned. Technician A says the engine cooling system should be checked for proper operation. Technician B says the customer may need an external cooler if using the vehicle for towing or performance applications. Who is right?
    A. A only
    B. B only
    C. Both A and B
    D. Neither A nor B (C.5)

47. Fluid sometimes escapes from the dipstick tube on an automatic transaxle. Technician A says the transaxle fluid may be contaminated. Technician B says the transaxle cooler may be defective. Who is right?
    A. A only
    B.` B only
    C. Both A and B
    D. Neither A nor B (A.1.3)

48. Technician A says case porosity between two fluid passages could result in pressure bleed off and clutch burnout. Technician B says case porosity between two fluid passages could result in the transmission being in two gears at once, resulting in a jam-up. Who is right?
    A. A only
    B. B only
    C. Both A and B
    D. Neither A nor B (D.2.8)

49. Technician A says on some transmissions, the throttle linkage may control both the downshift valve and the throttle valve. Technician B says some transmissions use a vacuum modulator to control the downshift valve. Who is right?
    A. A only
    B. B only
    C. Both A and B
    D. Neither A nor B (B.2)

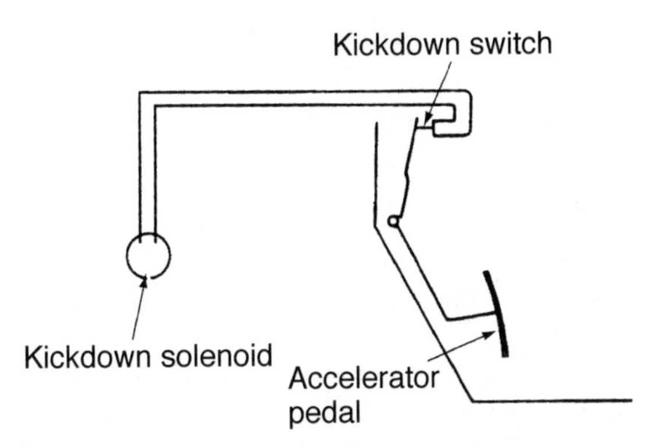

50. While discussing the proper way to diagnose a kickdown switch as shown in the figure above, Technician A says when the throttle pedal is fully depressed, a click should be heard just before the pedal reaches its travel stop. Technician B says if the transmission cannot be forced to automatically downshift, the kickdown switch could be open and should be replaced. Who is right?
    A. A only
    B. B only
    C. Both A and B
    D. Neither A nor B (B.2)
51. During a stall test, the stall speed is above specifications. Technician A says the exhaust system may be restricted. Technician B says the engine may have low compression. Who is right?
    A. A only
    B. B only
    C. Both A and B
    D. Neither A nor B (A.1.5)
52. A rear-wheel drive vehicle has a vibration that increases in relation to the vehicle speed. This vibration also is present when the engine is accelerated with the vehicle stopped and the gear selector in Neutral or Park. The cause of this problem could be:
    A. bad torque converter balance.
    B. bad driveshaft balance.
    C. bad driveshaft angles.
    D. bad engine mounts. (A.1.2)
53. A computer-controlled transaxle remains in second gear at all forward vehicle speed. The cause of the problem may be:
    A. a worn oil pump.
    B. a defective shift solenoid.
    C. a restricted filter.
    D. an improper linkage adjustment. (C.12)
54. Technician A says torque converter clutch (TCC) control problems are always caused by electrical malfunctions. Technician B says nearly all torque converter clutches (TCC) are engaged through the application of hydraulic pressure on the clutch. Who is right?
    A. A only
    B. B only
    C. Both A and B
    D. Neither A nor B (A.2.3)

55. Input and output shafts can be of the following types EXCEPT:
    A. solid.
    B. compound.
    C. tubular.
    D drilled. (D.2.4)

56. Technician A says an air test can be used to check servo action. Technician B says an air test can be used to check for internal fluid leaks. Who is right?
    A. A only
    B. B only
    C. Both A and B
    D. Neither A nor B (D.3.3)

57. While inspecting a drive chain and sprockets during a rebuild, the chain deflection is found to exceed specified limits. Technician A says that the sprockets should be checked for tooth wear. Technician B says that the drive chain can be shortened to repair the problem. Who is right?
    A. A only
    B. B only
    C. Both A and B
    D. Neither A nor B (D.2.9)

58. Technician A says throttle position is an important input for most electronic shift control systems. Technician B says vehicle speed is an important input for most electronic shift control systems. Who is right?
    A. A only
    B. B only
    C. Both A and B
    D. Neither A nor B (A.2.4)

59. A front-wheel drive vehicle experiences intermittent shifting. Sometimes the transaxle shifts normally, and occasionally it misses a shift. Technician A says the manual valve shift linkage may need adjusting. Technician B says the engine or transaxle mounts may be broken. Who is right?
    A. A only
    B. B only
    C. Both A and B
    D. Neither A nor B (C.13)

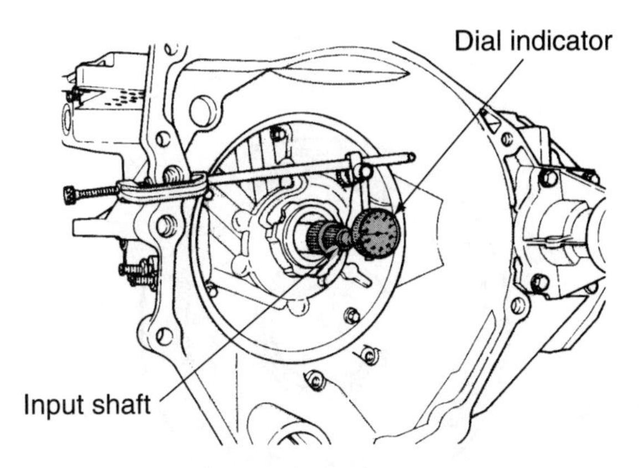

60. In the figure above if the input shaft end play is more than specified the most likely repair is:
   A. the input shaft must be replaced.
   B. the pump must be replaced.
   C. the transmission case is worn.
   D. a thicker thrust washer is required. (D1.2)

61. Technician A says all parts of the valve body should be soaked in mineral spirits before reassembling. Technician B says a lint-free rag must be used when wiping down valves. Who is right?
   A. A only
   B. B only
   C. Both A and B
   D. Neither A nor B (C.7)

62. An automatic transaxle has low pressure in third gear only. Technician A says the transaxle may have an internal leak. Technician B says the Throttle Valve (TV) cable may be misadjusted. Who is right?
   A. A only
   B. B only
   C. Both A and B
   D. Neither A nor B (A.1.4)

63. An automatic transmission has a whining noise that occurs in all gears while driving the vehicle. This noise is also present with the engine running and the vehicle stopped. Technician A says the rear planetary gearset may be defective. Technician B says the oil pump may be defective. Who is right?
   A. A only
   B. B only
   C. Both A and B
   D. Neither A nor B (A.1.2)

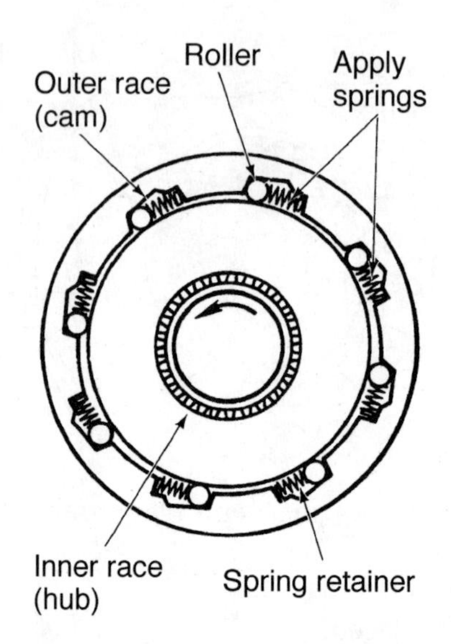

64. Technician A says a one-way overrunning clutch, shown above, can be either roller or sprag type. Technician B says a roller clutch, shown above, utilizes roller bearings held in place by sprags, which separate the inner and outer race of the clutch assembly. Who is right?
    A. A only
    B. B only
    C. Both A and B
    D. Neither A nor B (D.3.4)

65. When diagnosing a noise problem with an automatic transmission where the noise is present in all gears except park and neutral, the most likely causes are as follows EXCEPT:
    A. drive chain.
    B. torque converter.
    C. oil pump.
    D. input shaft. (D.1.2)

66. The engine shudders immediately after torque converter clutch (TCC) lockup. Technician A says the engine may have an ignition defect. Technician B says the fuel injection system may have a lean condition. Who is right?
    A. A only
    B. B only
    C. Both A and B
    D. Neither A nor B (A.1.6)

67. Technician A says that to road test an electronic transmission, you should connect a scan tool to monitor engine and transmission operation. Technician B says the electrical system should be checked for proper operation. Who is right?
    A. A only
    B. B only
    C. Both A and B
    D. Neither A nor B (A.2.1, A.2.5)

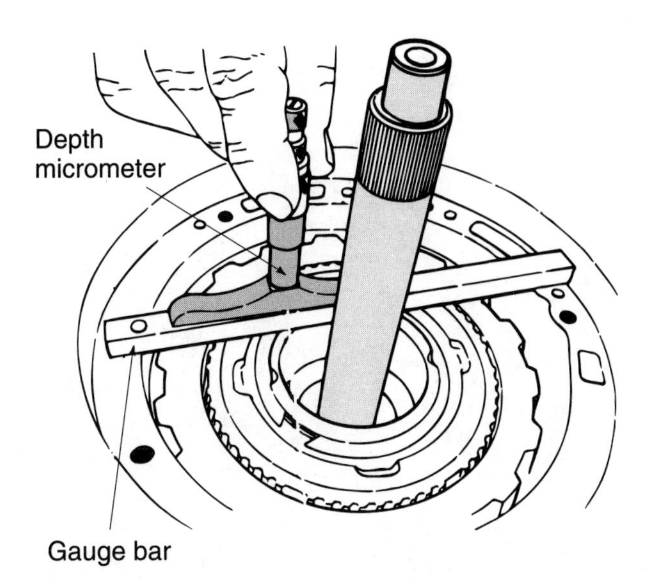

68. End play is being adjusted. The purpose of measuring the reverse clutch drum face with a depth micrometer, as shown, is to determine the proper:
    A. clutch pack retaining ring thickness.
    B. clutch pack reaction plate thickness.
    C. steel clutch plate thickness.
    D. selective washer thickness. (D.2.3)

6 Additional Test Questions for Practice

Additional Test Questions

Please note the letter and number in parentheses following each question. They match the overview in section 4 that discusses the relevant subject matter. You may want to refer to the overview using this cross-referencing key to help with questions posing problems for you.

1. Technician A says any worn or questionable parts involving the parking pawl should be replaced because of the safety risk. Technician B says the parking pawl can be inspected on some transmissions while it is still in the vehicle. Who is right?
   A. A only
   B. B only
   C. Both A and B
   D. Neither A nor B (C.11)
2. Technician A says that an inch-pound torque wrench is often used to measure preload. Technician B says that a digital micrometer is usually used to measure end-play. Who is right?
   A. A only
   B. B only
   C. Both A and B
   D. Neither A nor B (D.2.2)
3. Technician A says that sprag clutches with scored faces indicates fluid starvation. Technician B says that sprag clutches are less prone to fluid starvation than roller clutches. Who is right?
   A. A only
   B. B only
   C. Both A and B
   D. Neither A nor B (D.3.4)
4. During an inspection of the speedometer drive gear, it is found to be damaged. Technician A says if the drive gear is damaged, the driven gear should be inspected for damage. Technician B says if the driven gear is damaged, the whole output shaft of the transmission may have to be changed. Who is right?
   A. A only
   B. B only
   C. Both A and B
   D. Neither A nor B (C.6)

5. When diagnosing a computer controlled transaxle, all of these problems can affect shift timing and quality, EXCEPT:
   A. low battery voltage.
   B. defective vehicle speed sensor.
   C. defective wheel speed sensor.
   D. broken ground strap from body to transmission. (C.12)
6. Technician A says scan tools are needed to retrieve codes on all models of cars.Technician B says all scan tools provide a historical report of the computer system. Who is right?
   A. A only
   B. B only
   C. Both A and B
   D. Neither A nor B (A.2.1)
7. A vehicle was run with loose flexplate to converter connecting bolts. What is the LEAST likely damage that will result?
   A. Stripped converter connecting bolts.
   B. Discoloration of the converter from overheating.
   C. Cracked flexplate.
   D. Damaged converter pilot. (D.1.4)
8. While removing scratches in a valve, Technician A uses a fine emery cloth to remove the scratch. Technician B uses a sandblaster or glass bead machine to polish the surface of the valve. Who is right?
   A. A only
   B. B only
   C. Both A and B
   D. Neither A nor B (C.7)
9. After disconnecting the vacuum modulator vacuum line, fluid is found in the line. Technician A says the engine positive crankcase ventilation (PCV) system must be malfunctioning and is causing engine oil to enter the vacuum line. Technician B says the vacuum modulator should be checked for a torn diaphragm that is leaking transmission fluid into the vacuum line. Who is right?
   A. A only
   B. B only
   C. Both A and B
   D. Neither A nor B (C.1)

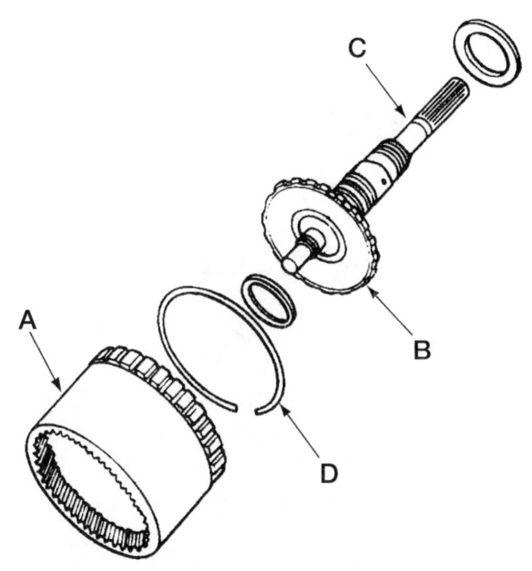

10. A parking pawl is found to have a damaged engagement surface. Which of the components shown in the figure above might also be damaged?
    A. A
    B. B
    C. C
    D. D (C.11)
11. Technician A says that before inspecting transmission parts they should be thoroughly cleaned. Technician B says that before disassembling a transmission endplay checks should be performed.Who is right?
    A. A only
    B. B only
    C. Both A and B
    D. Neither A nor B (D.1.2)
12. Technician A says the seals should be kept clean and free of dirt before and during installation. Technician B says most seals should be installed dry. Who is right?
    A. A only
    B. B only
    C. Both A and B
    D. Neither A nor B (D.2.5)
13. Technician A says that after verifying the customer complaint, the first thing to check on an automatic transmission is the level and condition of the fluid. Technician B says that a worn vacuum modulator can cause low fluid levels. Who is right?
    A. A only
    B. B only
    C. Both A and B
    D. Neither A nor B (A.1.1)
14. Technician A says that an excessively stretched chain in an automatic transaxle can cause harsh engagements. Technician B says that measuring its deflection with a dial indicator checks chain stretch. Who is right?
    A. A only
    B. B only
    C. Both A and B
    D. Neither A nor B (D.2.9)

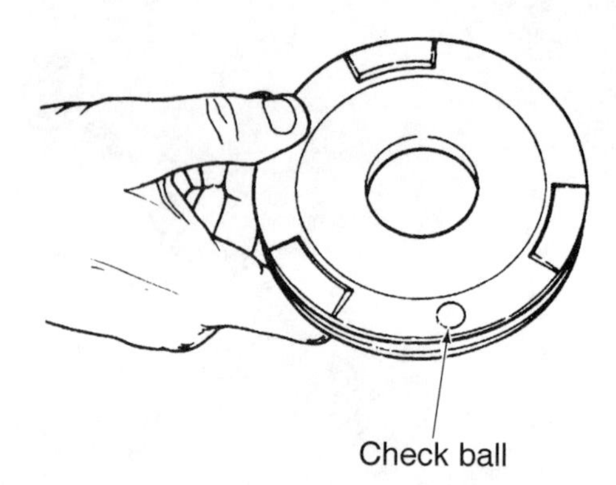

15. Refer to the above figure. What sound will be heard during an air pressure test if the encapsulated check ball in the in this clutch piston is missing?
    A. The compression of the clutch piston return spring.
    B. Clutch packs are not air pressure tested.
    C. A solid clunk.
    D. A hissing sound. (D.3.3)
16. While checking a transmission's vent, Technician A applies a vacuum to it and says that if the vent leaks, it should be replaced. Technician B runs automatic transmission fluid (ATF) through the vent and says that if the vent cannot hold fluid, it must be replaced. Who is right?
    A. A only
    B. B only
    C. Both A and B
    D. Neither A nor B (D.2.8)

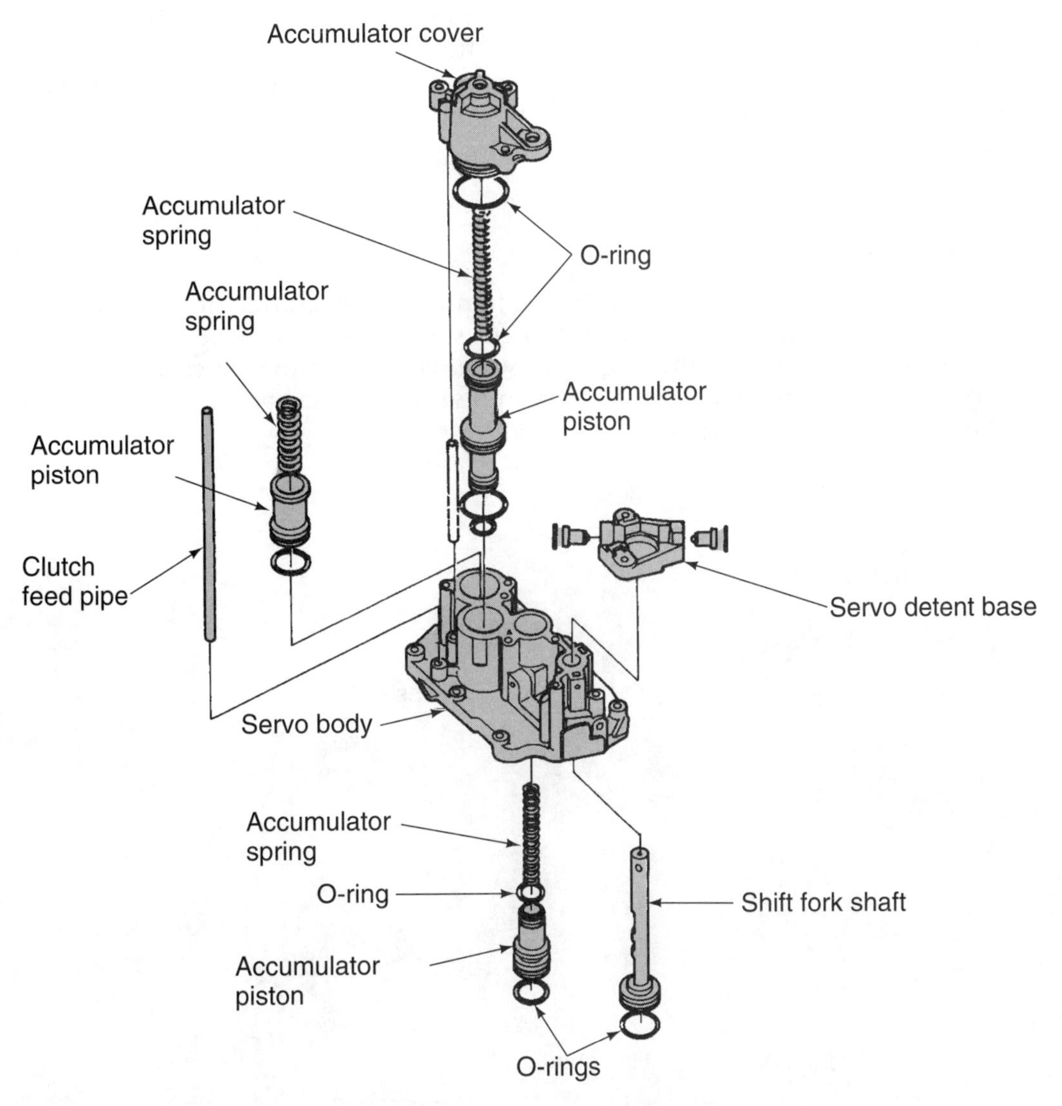

17. A servo body is shown in the figure above. Technician A says that if the accumulator piston is installed upside down, it will affect shift feel. Technician B says to always refer to a clear picture when you need to reinstall the accumulator. Who is right?
    A. A only
    B. B only
    C. Both A and B
    D. Neither A nor B (C.10)

18. What is the most likely result of improper transmission end play?
    A. Transmission case wear
    B. Input shaft bearing failure
    C. Delayed 1-2 uptight
    D. Driveling vibration (D.1.2)

19. Technician A says that flushing the transmission cooler reduces the chances that a rebuilt transmission will be contaminated after installation. Technician B says that compression fittings can be used to splice together metal cooler lines that were cut during transmission removal. Who is right?
    A. A only
    B. B only
    C. Both A and B
    D. Neither A nor B (C.5)

20. While diagnosing a computer controlled transaxle, a diagnostic trouble code representing the torque converter clutch (TCC) solenoid is obtained on a scan tester. Technician A says one of the torque converter clutch solenoid wires may be grounded inside the transaxle. Technician B says there may be an open wire in the transaxle electrical connector. Who is right?
    A. A only
    B. B only
    C. Both A and B
    D. Neither A nor B (A.2.3)

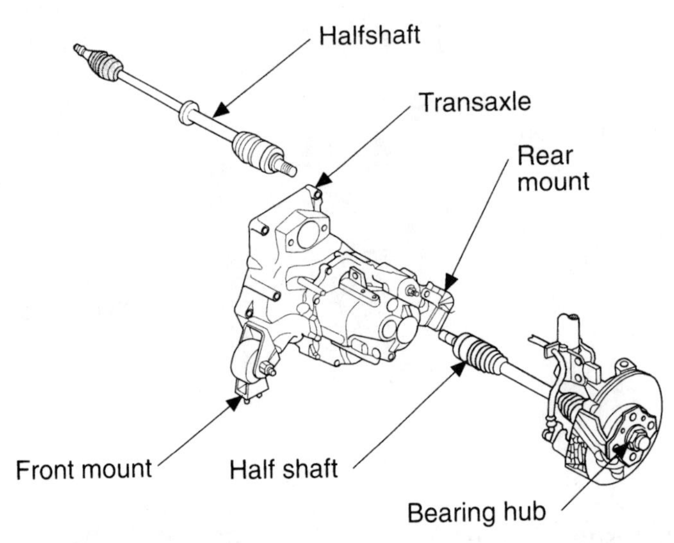

21. While diagnosing noises apparently coming from the transaxle assembly, shown above, Technician A says a knocking sound at all speeds with the wheels straight is probably caused by worn constant velocity (CV) joints. Technician B says a clicking noise heard when the vehicle is turning is probably caused by a worn or damaged outboard constant velocity (CV) joint. Who is right?
    A. A only
    B. B only
    C. Both A and B
    D. Neither A nor B (A.1.2)

22. Technician A says that a battery that has greater than 8.5 volts at the completion of a battery capacity test is good. Technician B says that a voltage drop test can be used to test most electrical circuits. Who is right?
    A. A only
    B. B only
    C. Both A and B
    D. Neither A nor B (A.2.5)

23. Technician A says that the gasket sometimes blocks valve body separator plate holes. Technician B says that overtightening a valve body can cause valve bores to distort. Who is right?
    A. A only
    B. B only
    C. Both A and B
    D. Neither A nor B (C.8)
24. A transmission case has cracks around several of the bell housing bolt holes. The most likely cause for this is:
    A. a defective transmission case.
    B. using a pry bar to separate the engine block and transmission after the bell housing bolts were removed.
    C. not marking the position of the flexplate in relation to the torque converter.
    D. the bell housing bolts were tightened with the case out of alignment. (D.1.1)
25. A deep groove is worn into the servo bore of a transmission case. What can be done to correct this condition?
    A. The servo bore can be honed, and an oversize piston installed.
    B. The servo bore can be honed, and a sleeve liner installed.
    C. The case must be replaced.
    D. Metal sealing rings can be used on the servo piston. (D.2.8)

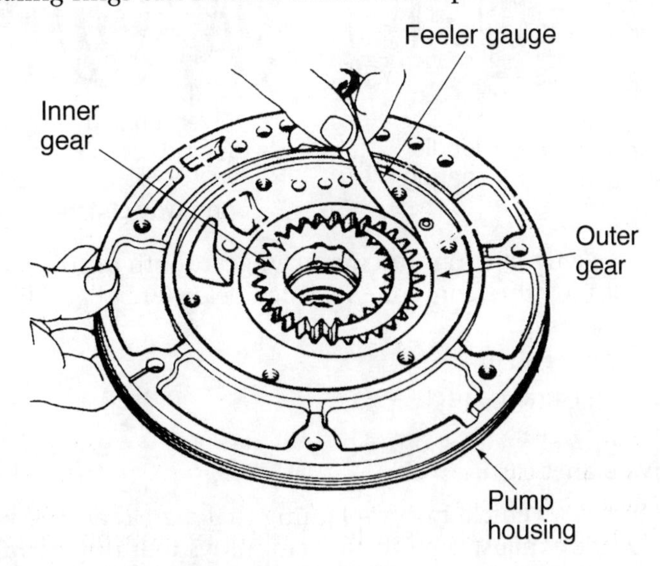

26. The Technician in the figure above is checking:
    A. pump gear-to-pocket clearance.
    B. pump gear pocket depth.
    C. pump gear-to-gear clearance.
    D. pump gear-to-surface angle. (D.2.1)
27. Technician A says that all transmissions use adjustable bands. Technician B says that band adjustment bolts should be tightened finger tight, then their lock nuts should be torqued to specifications. Who is right?
    A. A only
    B. B only
    C. Both A and B
    D. Neither A nor B (B.3)

28. Throttle linkage to a transmission performs all of the following EXCEPT:
    A. modifies throttle pressure.
    B. modifies governor pressure.
    C. modifies shift points.
    D. indicates engine load. (B.2)

29. Technician A says a typical torque converter clutch assembly's computer controls the application of the clutch by providing a ground circuit for the clutch solenoid circuit. Technician B says the clutch solenoid simply redirects fluid flow to activate the clutch.Who is right?
    A. A only
    B. B only
    C. Both A and B
    D. Neither A nor B (A.1.6)

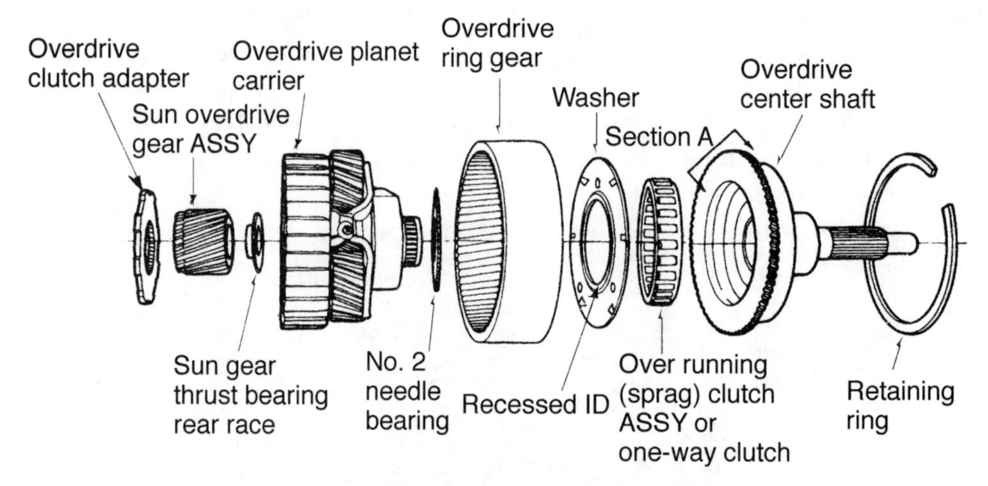

30. Refer to the above figure. You find a broken gear tooth on the sun overdrive gear ASSY. Which of the other components listed below are LEAST likely to require replacement as a result?
    A. Overdrive center shaft
    B. Overrunning (sprag) clutch ASSY
    C. Overdrive ring gear
    D. Overdrive planet carrier (D.2.7)

31. A customer brings a vehicle in complaining that a great deal of white smoke is coming out of the exhaust pipe.Technician A says that the modulator diaphragm may be leaking. Technician B says that fluid leaking from the modulator diaphragm may cause the vacuum hoses to become soft and spongy. Who is right?
    A. A only
    B. B only
    C. Both A and B
    D. Neither A nor B (A.1.1)

32. Technician A says that clutch discs do not need to be soaked in transmission fluid before assembly. Technician B says that dust, dirt or foreign matter on the parts during assembly won't affect transmission performance. Who is right?
    A. A only
    B. B only
    C. Both A and B
    D. Neither A nor B (D.3.1)

33. Technician A says that a pressure gauge is used to correctly adjust throttle linkage on some automatic transmissions. Technician B says that some manual linkage adjustments are done using a pressure gauge. Who is right?
    A. A only
    B. B only
    C. Both A and B
    D. Neither A nor B (B.1)

34. While discussing proper band adjustment procedures, Technician A says on some vehicles, the bands can be adjusted externally with a torque wrench. Technician B says a calibrated inch pound torque wrench is normally used to tighten the band adjusting the bolt to a specified torque. Who is right?
    A. A only
    B. B only
    C. Both A and B
    D. Neither A nor B (B.3)

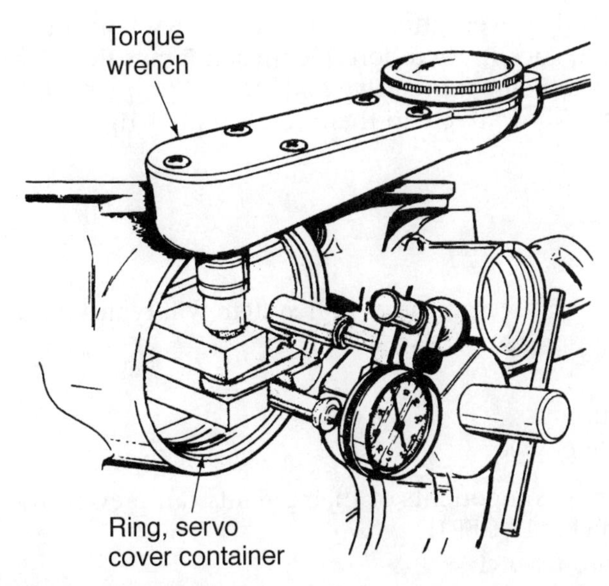

35. Technician A says that if you squeeze a band between your fingers and fluid comes out, the band should be replaced. Technician B says that strap or flex bands should never be flattened out when they are inspected. Who is right?
    A. A only
    B. B only
    C. Both A and B
    D. Neither A nor B (D.3.5)

36. Technician A says that to change the clutch pack clearance, you use varying thickness of snap rings. Technician B says that you can also vary the thickness of the clutch pressure plate. Who is right?
    A. A only
    B. B only
    C. Both A and B
    D. Neither A nor B (D.3.2)

37. Engine stalling at stops sometimes occurs on a TCC equipped car after the vehicle is decelerated from 55 mph (88.5 km/h). Technician A says the throttle position sensor (TPS) may be defective. Technician B says the brake switch may be defective. Who is right?
    A. A only
    B. B only
    C. Both A and B
    D. Neither A nor B (A.1.6)
38. Technician A says that transmission temperature sensors change resistance as the temperature of the transmission fluid changes. Technician B says that transmission temperature sensors can affect transmission shift speeds. Who is right?
    A. A only
    B. B only
    C. Both A and B
    D. Neither A nor B (A.2.4)
39. Technician A says that overtightening the hold-down bolts of the valve body can cause the valves to stick in their bore.Technician B says that flat filing the surface of the valve body will allow the valve body to seal properly and will therefore allow the valves to move freely in their bores. Who is right?
    A. A only
    B. B only
    C. Both A and B
    D. Neither A nor B (C.8)
40. A faulty governor or governor drive gear system will typically cause:
    A. improper shift points.
    B. harsh shifts.
    C. slipping shifts.
    D. excessive stall speeds. (C.2)
41. All of the following components or their circuits can prevent the torque converter clutch from applying EXCEPT:
    A. the brake on/off switch.
    B. the engine coolant temperature sensor.
    C. the throttle position sensor.
    D. the intake air temperature sensor. (A.2.3)
42. While inspecting an automatic transmission during teardown, the friction discs have a black line around the center of the friction surface. Technician A says that this condition is normal and no replacement is necessary. Technician B says that if a friction disc is squeezed and fluid comes to the surface, the disc is not glazed. Who is right?
    A. A only
    B. B only
    C. Both A and B
    D. Neither A nor B (D.3.1)
43. During a road test on a vehicle with an automatic transmission and lockup torque converter, a shudder is evident after converter lockup. All of the following could be the cause EXCEPT:
    A. worn universal joints.
    B. slipping overdrive clutch.
    C. defective torque converter clutch (TCC) solenoid.
    D. engine misfire. (A.1.6)

44. During deceleration a vehicle's torque converter clutch fails to disengage. Technician A says that a defective brake switch could cause this. Technician B says that a defective throttle position sensor could cause it. Who is right?
    A. A only
    B. B only
    C. Both A and B
    D. Neither A nor B (A.1.7)
45. Technician A says that in a vane type pump, the vane ring contains several sliding vanes that seal against a slide mounted in the pump housing. Technician B says vane type pumps are variable displacement pumps whose output can be reduced when high fluid pressures are not necessary. Who is right?
    A. A only
    B. B only
    C. Both A and B
    D. Neither A nor B (D.2.1)
46. Technician A says that a converter clutch that fails to apply can cause a decrease in the vehicle's fuel economy. Technician B says that a converter clutch that fails to apply can cause overheating. Who is right?
    A. A only
    B. B only
    C. Both A and B
    D. Neither A nor B (A.1.6)
47. During a pressure test, a transmission has high pressure in all gears and ranges. This may be caused by:
    A. a stuck pressure regulator valve.
    B. a damaged pump.
    C. a broken manual valve spring.
    D. a broken accumulator spring. (A.1.4)

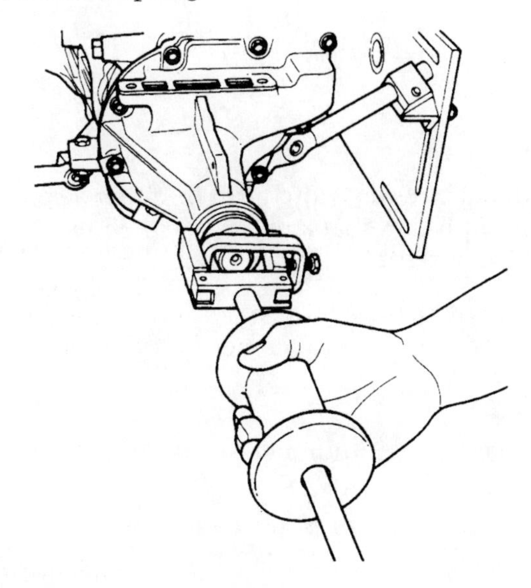

48. The Technician in the figure above is:
    A. installing an extension housing seal.
    B. removing an extension housing bushing.
    C. removing an extension housing seal.
    D. installing an extension housing bushing. (C.4)

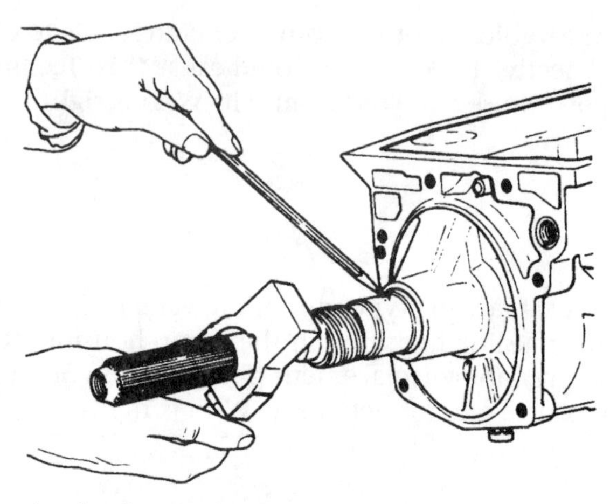

49. The Technician in the figure above is:
    A. removing or installing a governor.
    B. removing or installing an output shaft using the correct special tools.
    C. replacing an output shaft bushing.
    D. removing a speedometer drive gear. (C.2)
50. Technician A says that a transmission filter may be cleaned and reused. Technician B says that blackish deposits in the bottom of the pan indicates a rebuild is necessary. Who is right?
    A. A only
    B. B only
    C. Both A and B
    D. Neither A nor B (B.4)
51. While inspecting a variable displacement vane type pump, the vanes show no cracks and have flat ends. Technician A says the vanes are acceptable. Technician B says that the vanes need to be replaced and sizing for replacement is not necessary. Who is right?
    A. A only
    B. B only
    C. Both A and B
    D. Neither A nor B (D.2.1)
52. Technician A says that a one-way clutch can be either a sprag type or a roller and spring type . Technician B says that a one-way clutch requires a special hydraulic circuit to operate. Who is right?
    A. A only
    B. B only
    C. Both A and B
    D. Neither A nor B (D.3.4)
53. If a one way clutch has failed, a thorough inspection of the:
    A. bands and servos must be made.
    B. transmission cooling system must be made.
    C. clutch pack assemblies must be made.
    D. hydraulic fluid feed circuit must be made. (D.3.4)

54. Technician A says a crack around a bolt hole indicates that the case bolts were tightened with the case out of alignment to the engine block. Technician B says the case should be replaced if cracked, and not repaired. Who is right?
    A. A only
    B. B only
    C. Both A and B
    D. Neither A nor B (D.1.1)

RANGE	GEAR	A SOLENOID	B SOLENOID	4TH CLUTCH	REVERSE BAND	2ND CLUTCH	3RD CLUTCH	3RD ROLLER CLUTCH	INPUT CLUTCH	INPUT SPRAG	FORWARD BAND	1/2 SUPPORT ROLLER CLUTCH	2/1 BAND
P-N		ON	ON						*	*			
D	1st	ON	ON						APPLIED	HOLDING	APPLIED	HOLDING	
	2nd	OFF	ON			APPLIED			*	OVERRUN	APPLIED	HOLDING	
	3rd	OFF	OFF			APPLIED	APPLIED	HOLDING			APPLIED	OVERRUN	
	4th	ON	OFF	APPLIED		APPLIED	*	OVERRUN			APPLIED	OVERRUN	
D	3rd	@OFF	@OFF			APPLIED	APPLIED	HOLDING	APPLIED	HOLDING	APPLIED	OVERRUN	
	2nd	@OFF	@ON			APPLIED			*	OVERRUN	APPLIED	HOLDING	
	1st	@ON	@ON						APPLIED	HOLDING	APPLIED	HOLDING	
2	2nd	@OFF	@ON			APPLIED			*	OVERRUN	APPLIED	HOLDING	APPLIED
	1st	@ON	@ON						APPLIED	HOLDING	APPLIED	HOLDING	APPLIED
1	1st	@ON	@ON				APPLIED	HOLDING	APPLIED	HOLDING	APPLIED	HOLDING	APPLIED
R	REVERSE	ON	ON		APPLIED				APPLIED	HOLDING			

*APPLIED BUT NOT EFFECTIVE

ON - SOLENOID ENERGIZED

OFF - SOLENOID DE-ENERGIZED

@ THE SOLENOID'S STATE FOLLOWS A SHIFT PATTERN WHICH DEPENDS UPON VEHICLE SPEED AND THROTTLE POSITION. IT DOES NOT DEPEND UPON THE SELECTED GEAR.

55. Refer to the apply chart above. A vehicle with this type of automatic transmission makes a noise on take-off in manual 1st that disappears when the shift selector is placed in automatic 1st. Technician A says the noise may be coming from the 3rd clutch. Technician B says the noise may be coming from the 2nd clutch. Who is right?
    A. A only
    B. B only
    C. Both A and B
    D. Neither A nor B (A.1.2)

56. While discussing the results of an oil pressure test, Technician A says when the fluid pressures are low, an internal leak, a clogged filter, a low oil pump output, or a faulty pressure regulator valve are indicated. Technician B says if the fluid pressure increases at the wrong time, an internal leak at the servo or clutch seal is indicated. Who is right?
    A. A only
    B. B only
    C. Both A and B
    D. Neither A nor B (A.1.4)

57. Air pressure tests can be used to check all of the following EXCEPT:
    A. clutch pack clearance.
    B. servo seal installation.
    C. clutch piston seal installation.
    D. hydraulic circuit passages for blockages. (D.3.3)

58. Technician A says that some vacuum modulators can be adjusted to change transmission shift points. Technician B says that adjustments are made to vacuum modulators using shims. Who is right?
    A. A only
    B. B only
    C. Both A and B
    D. Neither A nor B (C.1)
59. Technician A says that some transmissions are only partially controlled electronically. Technician B says that some electronic transmissions have governors. Who is right?
    A. A only
    B. B only
    C. Both A and B
    D. Neither A nor B (C.12)
60. Technician A says that some transmissions have mated accumulator and servo pistons. Technician B says that because of their design, accumulator pistons cannot be installed upside down. Who is right?
    A. A only
    B. B only
    C. Both A and B
    D. Neither A nor B (C.10)
61. All of the following can result in damage to a one-way clutch or planetary gearset EXCEPT:
    A. a check ball that correctly seats in a shaft oil passage.
    B. a missing check ball in a shaft oil passage.
    C. an obstructed oil passage in a shaft.
    D. a worn shaft bushing. (D.2.4)
62. Technician A says that replacement speedometer gears must have the same number of teeth as the original. Technician B says that incorrect speed sensor readings will not affect transmission operation. Who is right?
    A. A only
    B. B only
    C. Both A and B
    D. Neither A nor B (C.6)
63. When measuring the clearance on a clutch pack with a wavy snap ring, you should measure from the top of the pressure plate to:
    A. the lowest point of the snap ring.
    B. a point between the highest and lowest point of the snap ring.
    C. the highest point of the snap ring.
    D. the highest point of the snap ring groove. (D.3.2)
64. Technician A says bushings should be heated with a torch to easily remove them. Technician B says bushings should be removed with a slide hammer and the correct attachment. Who is right?
    A. A only
    B. B only
    C. Both A and B
    D. Neither A nor B (D.2.6)

65. Bushings can be removed by using:
    A. a slide hammer.
    B. heat.
    C. chisel.
    D. snap ring pliers. (D.2.6)
66. Thrust washers can have all of the following EXCEPT:
    A. selective thickness.
    B. holes to allow oil to pass.
    C. positioning tabs to help hold them in place.
    D. selective inner diameters. (D.2.3)

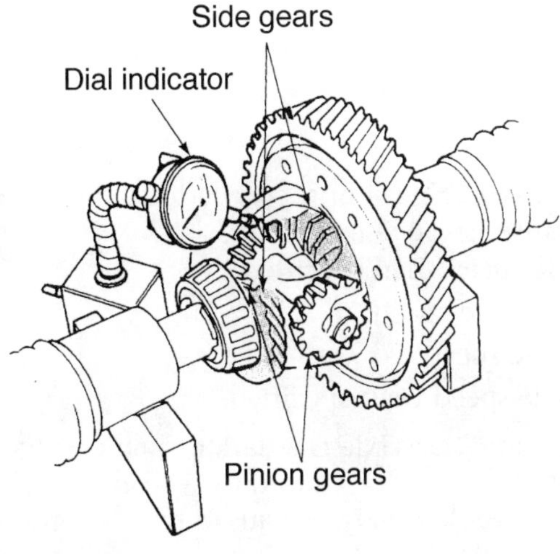

67. In the figure above, what measurement is being taken?
    A. Side gear end-play
    B. Side gear turning torque
    C. Differential end-play
    D. Side gear backlash (D.2.10)
68. Technician A says that when installing lip seals on the clutch, the lip should always face the clutch pressure plate. Technician B says that if the steel clutch plates pass inspection, remove the polished surface finish before reuse. Who is right?
    A. A only
    B. B only
    C. Both A and B
    D. Neither A nor B (D.3.1)
69. Technician A says that during transmission assembly, bushings are usually held in place with grease. Technician B says that the depth of installation and oil groove position should be noted before removal of bushings. Who is right?
    A. A only
    B. B only
    C. Both A and B
    D. Neither A nor B (D.2.6)

70. Technician A says that damaged transmission mounts on a rear wheel drive vehicle can cause uneven tire wear. Technician B says that damaged transmission mounts can cause vibrations even when the transmission is in neutral. Who is right?
    A. A only
    B. B only
    C. Both A and B
    D. Neither A nor B (C.13)
71. An oil leak is found coming from the extension housing tail seal. Technician A says it could be caused by a bad extension housing bushing. Technician B says to check the clearance between the sliding yoke and the bushing. Who is right?
    A. A only
    B. B only
    C. Both A and B
    D. Neither A nor B (C.4)
72. When the throttle valve cable is improperly adjusted so throttle pressure is higher than normal, the transmission shifts occur:
    A. at a lower vehicle speed than specified.
    B. at the specified vehicle speed.
    C. at the same vehicle speed.
    D. at a higher vehicle speed than specified. (B.2)
73. The fluid in an automatic transaxle is a dark brown color and smells burned.Technician A says this problem may be caused by a worn front planetary sun gear. Technician B says the problem may be caused by worn friction-type clutch plates. Who is right?
    A. A only
    B. B only
    C. Both A and B
    D. Neither A nor B (A.1.3)
74. When performing a stall test, the throttle should be held wide open for:
    A. 3 seconds.
    B. 5 seconds.
    C. 7 seconds.
    D. 9 seconds. (A.1.5)
75. Technician A says all types of automatic transmission fluid (ATF) become lighter when the friction modifiers are depleted. Technician B says petroleum based automatic transmission fluid (ATF) typically has a clear, red color and will darken when it is burned. Who is right?
    A. A only
    B. B only
    C. Both A and B
    D. Neither A nor B (C.5)

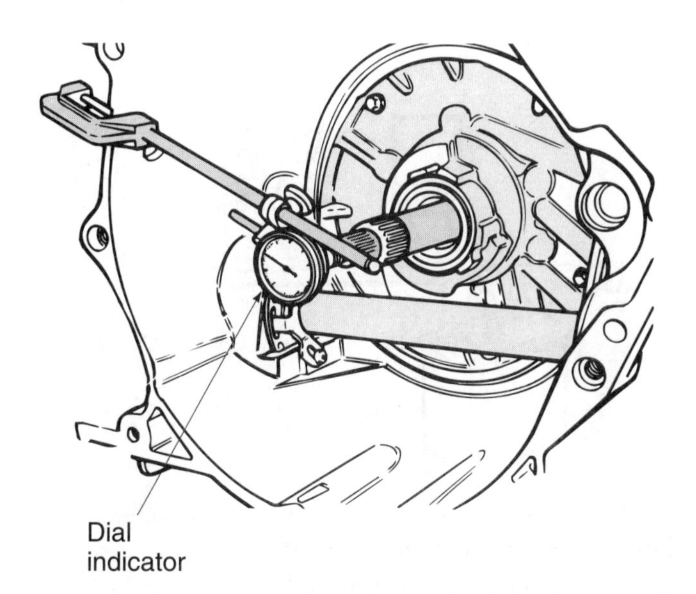

76. While discussing shaft end play, as shown above, Technician A says shaft end play is measured only after rebuilding. Technician B says shaft end play can be corrected by the addition or subtraction of shims. Who is right?
    A. A only
    B. B only
    C. Both A and B
    D. Neither A nor B (D.2.3)
77. While discussing the levers that protrude from the transmission housing, Technician A says the throttle lever is moved by the engine vacuum sent to the modulator in response to engine load. Technician B says that when the gear shift selector is moved, it moves the manual lever, which moves the manual shift valve in the transmission's valve body. Who is right?
    A. A only
    B. B only
    C. Both A and B
    D. Neither A nor B (B.1)

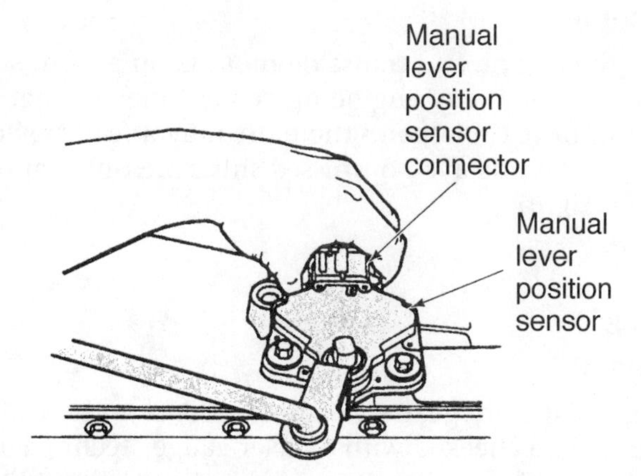

78. A torque converter should be replaced if any of the following signs are evident, EXCEPT:
    A. loose drive studs.
    B. light scratches on the drive hub.
    C. leaking at seams or welds.
    D. excessive runout. (D.1.4)

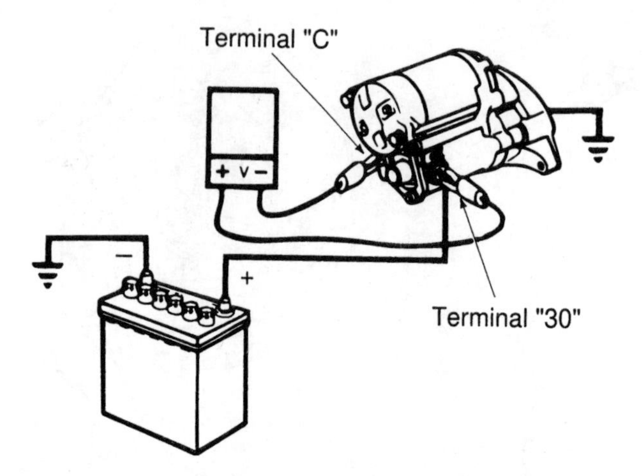

79. The digital volt/ohmmeter (DVOM) in the figure above is connected to test:
    A. voltage drop across the solenoid contacts.
    B. voltage drop on the positive battery cable.
    C. resistance of the positive battery cable.
    D. voltage drop across the starter motor. (D.2.5)
80. Technician A says before proceeding with assembly of the transmission, it is important to flush the case with automatic transmission fluid (ATF). Technician B says to use vaseline for securing washers during installation. Who is right?
    A. A only
    B. B only
    C. Both A and B
    D. Neither A nor B (D.1.3)
81. An automatic transmission-equipped vehicle is brought in low on fluid. Technician A says that a pressure test should be the first diagnostic procedure performed. Technician B says that this could be the result of a leaking speedometer seal O-ring. Who is right?
    A. A only
    B. B only
    C. Both A and B
    D. Neither A nor B (C.3)
82. While checking the engine and transaxle mounts on a front wheel drive (FWD) vehicle, Technician A says any engine movement may change the effective length of the shift and throttle cables, and therefore may affect the engagement of the gears. Technician B says delayed or missed shifts are only caused by hydraulic problems. Who is right?
    A. A only
    B. B only
    C. Both A and B
    D. Neither A nor B (C.13)
83. While checking a planetary gearset, Technician A says the end clearance of the pinion gears should be checked with a feeler gauge. Technician B says the end clearance of the long pinions in a Ravigneaux gearset should be at both ends. Who is right?
    A. A only
    B. B only
    C. Both A and B
    D. Neither A nor B (D.2.7)

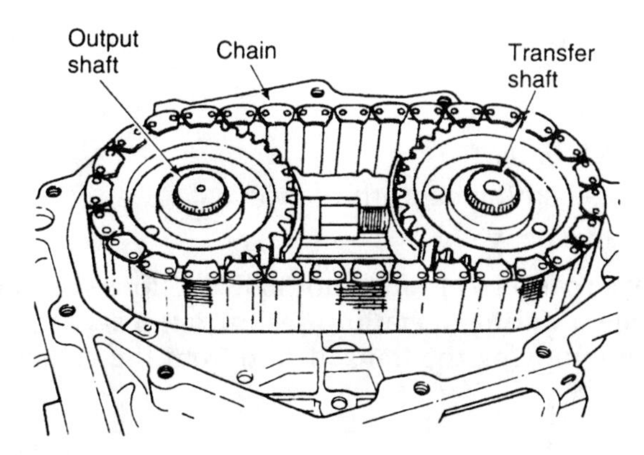

84. In the figure above, Technician A says that drive chains are usually not replaced until they break. Technician B says to mark the drive chain before removing it in order to indicate which side was up. Who is right?
    A. A only
    B. B only
    C. Both A and B
    D. Neither A nor B (D.2.9)
85. When performing a pressure test, Technician A says to use high pressure. Technician B says air leaks at the Teflon™ seal area are normal. Who is right?
    A. A only
    B. B only
    C. Both A and B
    D. Neither A nor B (D.3.3)
86. A transaxle experiences premature band failure, and the band is worn severely on the outer edges, but not in the center. Technician A says the band strut may be worn unevenly, resulting in band misalignment. Technician B says the band contact area on the clutch drum is dished. Who is right?
    A. A only
    B. B only
    C. Both A and B
    D. Neither A nor B (D.3.5)
87. The transmission fluid is found to be pink and milky. This is caused by:
    A. overheating of the fluid.
    B. engine oil mixing with the fluid.
    C. engine coolant mixing with the fluid.
    D. aeration of the fluid. (A.1.3)
88. A customer is complaining about harsh engagements and hard shifting of their electronic automatic transmission. This can be caused by:
    A. an open circuit to the electronic pressure control solenoid.
    B. an open circuit to a shift solenoid.
    C. a short to power in the electronic pressure control solenoid control circuit.
    D. a short to power in a shift solenoid control circuit. (A.2.1)
89. Technician A says that any contamination or black deposits found in the transmission pan indicate serious internal damage. Technician B says that most transmission filters can be cleaned and reused. Who is right?
    A. A only
    B. B only
    C. Both A and B
    D. Neither A nor B (B.4)

90. When servicing a valve body, if a valve cannot be made to move freely in its bore, you should:
    A. replace the valve.
    B. hone the valve's bore.
    C. install a stronger spring beneath the valve.
    D. replace the valve body. (C.7)
91. Technician A says that servo pistons normally are loose on their guide pins to allow for movement during operation. Technician B says that some servos can be replaced without removing the transmission from the vehicle. Who is right?
    A. A only
    B. B only
    C. Both A and B
    D. Neither A nor B (C.9)
92. During a transmission pressure test, the pressure gradually decreases at higher engine speeds. The cause of these problem could be:
    A. a worn oil pump.
    B. a restricted oil filter.
    C. a stuck pressure regulator.
    D. a plugged modulator hose. (A.1.4)
93. During a stall test, the engine RPM is lower than normal. Technician A says that a slipping torque converter clutch may be the cause. Technician B says an engine performance problem may be the cause. Who is right?
    A. A only
    B. B only
    C. Both A and B
    D. Neither A nor B (A.1.5)

Appendices

Answers to the Test Questions for the Sample Test Section 5

1.	C	18.	C	35.	D	52.	A
2.	D	19.	B	36.	A	53.	B
3.	A	20.	C	37.	B	54.	B
4.	B	21.	A	38.	C	55.	B
5.	B	22.	D	39.	B	56.	A
6.	D	23.	C	40.	C	57.	A
7.	A	24.	D	41.	B	58.	C
8.	C	25.	D	42.	D	59.	B
9.	C	26.	A	43.	C	60.	D
10.	C	27.	B	44.	B	61.	A
11.	D	28.	C	45.	A	62.	A
12.	B	29.	A	46.	C	63.	B
13.	C	30.	A	47.	D	64.	D
14.	D	31.	C	48.	C	65.	C
15.	A	32.	D	49.	A	66.	C
16.	A	33.	A	50.	C	67.	C
17.	B	34.	A	51.	D	68.	D

Explanations to the Answers for the Sample Test Section 5

Question #1
Answer A is wrong.
Answer B is wrong.
Answer C is correct.
Answer D is wrong.

Question #2
Answer A is wrong. Scoring occurs when there is a lack of lubrication.
Answer B is wrong. Wire-type feeler gauges can be used to check shaft-to-bushing clearance.
Answer C is wrong. Shaft-to-bushing clearance can also be checked with a vernier caliper and a micrometer.
Answer D is correct.

Question #3
Answer A is correct.
Answer B is wrong. A sticking pressure regulator would cause all shifts to be harsh, not just one.
Answer C is wrong.
Answer D is wrong.

Question #4
Answer A is wrong. Transmission cooler lines are removed before the transmission.
Answer B is correct.
Answer C is wrong. The inspection cover is removed before the transmission.
Answer D is wrong. The driveshaft is removed before the transmission.

Question #5
Answer A is wrong. Steel plates should be worn flat, but should be checked for thickness; plates that are too thin should be replaced.
Answer B is correct.
Answer C is wrong.
Answer D is wrong.

Question #6
Answer A is wrong. A sticking clutch drum check ball can cause the clutch discs to slip.
Answer B is wrong. A reduced clutch pack clearance can cause excessive friction on the clutch discs.
Answer C is wrong. A damaged clutch piston seal can cause a fluid leak and a pressure drop.
Answer D is correct.

Question #7
Answer A is correct.
Answer B is wrong. Blackened gears are indications of overloading and the carrier should be replaced.
Answer C is wrong.
Answer D is wrong.

Question #8
Answer A is wrong.
Answer B is wrong. If the drainback hole behind the seal is plugged, the pressure would blow the seal out, not just leak.
Answer C is correct.
Answer D is wrong.

Question #9
Answer A is wrong.
Answer B is wrong.
Answer C is correct.
Answer D is wrong.

Question #10
Answer A is wrong.
Answer B is wrong.
Answer C is correct.
Answer D is wrong.

Question #11
Answer A is wrong. A leaking converter drain plug would cause the front of the converter to be wet with transmission fluid.
Answer B is wrong. A leaking rear main would cause the front of the converter to be wet with oil.
Answer C is wrong. A loose rear main bearing would cause a noisy engine, and any leak would cause the front of the converter to be wet with oil.
Answer D is correct.

Question #12
Answer A is wrong. A plugged or restricted transmission vent would cause excessive pressures and an external oil leak.
Answer B is correct.
Answer C is wrong.
Answer D is wrong.

Question #13
Answer A is wrong.
Answer B is wrong.
Answer C is correct.
Answer D is wrong.

Question #14
Answer A is wrong. A defective stator one-way clutch would cause low engine stall speed, not a defective turbine.
Answer B is wrong. Slipping transmission clutches cause higher engine stall speeds, not lower.
Answer C is wrong.
Answer D is correct.

Question #15
Answer A is correct.
Answer B is wrong. If the problem was not present before the rebuild, it is most likely that the valve body was over-torqued.
Answer C is wrong.
Answer D is wrong.

Question #16
Answer A is correct.
Answer B is wrong. An air check cannot be used to check for external leaks.
Answer C is wrong.
Answer D is wrong.

Question #17
Answer A is wrong. Potentiometers are used to measure movement like a throttle position sensor.
Answer B is correct.
Answer C is wrong.
Answer D is wrong.

Question #18
Answer A is wrong.
Answer B is wrong.
Answer C is correct.
Answer D is wrong.

Question #19
Answer A is wrong. Most transmission problems can be identified without conducting a pressure test.
Answer B is correct.
Answer C is wrong.
Answer D is wrong.

Question #20
Answer A is wrong.
Answer B is wrong.
Answer C is correct.
Answer D is wrong.

Question #21
Answer A is correct.
Answer B is wrong. An accumulator is a springloaded piston that provides a hydraulic cushion to reduce shift harshness.
Answer C is wrong.
Answer D is wrong.

Question #22
Answer A is wrong. Poor gear engagement is a likely result of the indicated problem.
Answer B is wrong. Low fluid pressure is a likely result of the indicated problem.
Answer C is wrong. Incorrect gear reading is a likely result of the indicated problem.
Answer D is correct.

Question #23
Answer A is wrong. There are no thrust washers in the clutch packs.
Answer B is wrong. The reaction plate should be replaced with one of the proper thickness.
Answer C is correct.
Answer D is wrong. Replacing a clutch drum will not correct the clearance problem.

Question #24
Answer A is wrong. The flow should be 1 quart in 20 seconds.
Answer B is wrong. The flow should be 1 quart in 20 seconds.
Answer C is wrong. The flow should be 1 quart in 20 seconds.
Answer D is correct.

Question #25
Answer A is wrong. The first step when removing the transmission is to disconnect the negative battery cable.
Answer B is wrong. For balancing purposes, always mark the driveshaft.
Answer C is wrong. Loosening the engine support fixture aids in removing the transaxle.
Answer D is correct.

Question #26
Answer A is correct.
Answer B is wrong. A dry speedometer cable would squeak or make an audible noise while the vehicle is in motion.
Answer C is wrong. The extra clearance between the gear tooth would cause erratic readings.
Answer D is wrong. A worn retaining bushing would cause the gear drive to "walk around", causing erratic readings.

Question #27
Answer A is wrong. The strategy allows for one forward gear and one gear in reverse, but not one in first gear.
Answer B is correct.
Answer C is wrong. The strategy will not provide two forward speeds.
Answer D is wrong. The strategy will not provide two forward speeds.

Question #28
Answer A is wrong.
Answer B is wrong.
Answer C is correct.
Answer D is wrong.

Question #29
Answer A is correct.
Answer B is wrong. A misadjusted shift linkage will cause lower than normal fluid pressure.
Answer C is wrong.
Answer D is wrong.

Question #30
Answer A is correct.
Answer B is wrong. The discharge rate is one-half the cold cranking amp rating of the battery.
Answer C is wrong.
Answer D is wrong.

Question #31
Answer A is wrong.
Answer B is wrong.
Answer C is correct.
Answer D is wrong.

Question #32
Answer A is wrong. This is a common leak source.
Answer B is wrong. This is a common leak source.
Answer C is wrong. This is a common leak source.
Answer D is correct.

Question #33
Answer A is correct.
Answer B is wrong. The parking pawl locks the output shaft.
Answer C is wrong.
Answer D is wrong.

Question #34
Answer A is correct.
Answer B is wrong. The valve body may not have to be removed.
Answer C is wrong.
Answer D is wrong.

Question #35
Answer A is wrong. Sticking governors are usually indicated by a no-shifting condition.
Answer B is wrong. Excessive governor tension will cause shifts at lower speeds.
Answer C is wrong. Worn governor weights or springs will cause erratic shifts at all speeds.
Answer D is correct.

Question #36
Answer A is correct.
Answer B is wrong. The clearance should not exceed the specified limit.
Answer C is wrong. The clearance should not exceed the specified limit.
Answer D is wrong. The clearance should not exceed the specified limit.

Question #37
Answer A is wrong. The vacuum modulator should have manifold vacuum.
Answer B is correct.
Answer C is wrong.
Answer D is wrong.

Question #38
Answer A is wrong.
Answer B is wrong.
Answer C is correct.
Answer D is wrong.

Question #39
Answer A is wrong. Main bearings would cause engine noise under a load and on acceleration, not deceleration.
Answer B is correct.
Answer C is wrong. Connecting rod bearings would cause engine noise under a load and on acceleration, not deceleration.
Answer D is wrong. Piston pins would cause engine noise under a load and on acceleration, not deceleration.

Question #40
Answer A is wrong.
Answer B is wrong.
Answer C is correct.
Answer D is wrong.

Question #41
Answer A is wrong. Adding thicker spaces to both side gears would not allow enough free-play.
Answer B is correct.
Answer C is wrong. Adding spacers to the bearing cups would not correct the torque measurement.
Answer D is wrong. Adding spacers behind the side bearing will only affect the side bearing preload.

Question #42
Answer A is wrong. Cables can dry up and fracture.
Answer B is wrong. Missing gear teeth can often cause speedometer readings to be incorrect.
Answer C is wrong. Speedometers have many small parts that can fail.
Answer D is correct.

Question #43
Answer A is wrong.
Answer B is wrong.
Answer C is correct.
Answer D is wrong.

Question #44
Answer A is wrong. Shift speeds are controlled by the throttle valve.
Answer B is correct.
Answer C is wrong. Shift speeds are controlled by the throttle valve.
Answer D is wrong. If slipping is in more than one gear, it cannot be a single band.

Question #45
Answer A is correct.
Answer B is wrong. A defective stator clutch does not affect converter clutch lockup.
Answer C is wrong. A second speed servo does not affect converter clutch lockup.
Answer D is wrong. A pressure regulator does not affect converter clutch lockup.

Question #46
Answer A is wrong.
Answer B is wrong.
Answer C is correct.
Answer D is wrong.

Question #47
Answer A is wrong. Fluid from the dipstick may be caused by aeration due to incorrect fluid level, not contamination.
Answer B is wrong. A defective transaxle cooler would cause fluid contamination.
Answer C is wrong.
Answer D is correct.

Question #48
Answer A is wrong.
Answer B is wrong.
Answer C is correct.
Answer D is wrong.

Question #49
Answer A is correct.
Answer B is wrong. Vacuum modulators are used to control the throttle valve (TV), not downshift valves.
Answer C is wrong.
Answer D is wrong.

Question #50
Answer A is wrong.
Answer B is wrong.
Answer C is correct.
Answer D is wrong.

Question #51
Answer A is wrong.
Answer B is wrong.
Answer C is wrong.
Answer D is correct.

Question #52
Answer A is correct.
Answer B is wrong. The driveshaft does not rotate when the vehicle is in park or neutral, and could not cause the vibration.
Answer C is wrong. A bad driveshaft angle will not affect the vehicle when it is in park or neutral, and could not cause the vibration.
Answer D is wrong. Bad engine mount will not cause vibrations when the vehicle is in park or neutral, and could not cause the vibration.

Question #53
Answer A is wrong. A worn oil pump would cause problems in every gear, but would not cause the transmission to remain in only one gear.
Answer B is correct.
Answer C is wrong. A restricted filter would cause problems in every gear, but would not cause the transmission to remain in only one gear.
Answer D is wrong. An improper linkage adjustment would allow other gears to be manually chosen even if the PRNDL indicator was not on the correct gear.

Question #54
Answer A is wrong. Poor lockup action can be caused by engine, electrical, clutch, or torque converter problems.
Answer B is correct.
Answer C is wrong.
Answer D is wrong.

Question #55
Answer A is wrong. Solid is a type mentioned in the text.
Answer B is correct.
Answer C is wrong. Tubular is a type mentioned in the text.
Answer D is wrong. Drilled is a type mentioned in the text.

Question #56
Answer A is correct.
Answer B is wrong. An air test is never used to find internal fluid leaks.
Answer C is wrong.
Answer D is wrong.

Question #57
Answer A is correct.
Answer B is wrong. The drive chain cannot be shortened as they are specific lengths and should be replaced when stretched.
Answer C is wrong.
Answer D is wrong.

Question #58
Answer A is wrong.
Answer B is wrong.
Answer C is correct.
Answer D is wrong.

Question #59
Answer A is wrong. If the manual valve shift linkage were misadjusted, it would cause a problem all of the time, not intermittently.
Answer B is correct.
Answer C is wrong.
Answer D is wrong.

Question #60
Answer A is wrong. Input shafts do not come in different lengths.
Answer B is wrong. There is no relationship between the pump and the input shaft end play.
Answer C is wrong. The transmission case is not related to the input shaft end play.
Answer D is correct.

Question #61
Answer A is correct.
Answer B is wrong. Compressed air should be used to dry parts; a rag will leave debris.
Answer C is wrong.
Answer D is wrong.

Question #62
Answer A is correct.
Answer B is wrong. A misadjusted throttle valve cable would cause low pressure in all gears, not just one gear.
Answer C is wrong.
Answer D is wrong.

Question #63
Answer A is wrong. Because the noise occurs when the vehicle is stopped and the planetary gearset is not in use when the vehicle is stopped, it could not be the cause of the problem.
Answer B is correct.
Answer C is wrong.
Answer D is wrong.

Question #64
Answer A is wrong. The figure shows a roller type one-way overrunning clutch, not a sprag type.
Answer B is wrong. The rollers are held in place by applying springs in the figure shown.
Answer C is wrong.
Answer D is correct.

Question #65
Answer A is wrong.
Answer B is wrong.
Answer C is correct.
Answer D is wrong.

Question #66
Answer A is wrong.
Answer B is wrong.
Answer C is correct.
Answer D is wrong.

Question #67
Answer A is wrong.
Answer B is wrong.
Answer C is correct.
Answer D is wrong.

Question #68
Answer A is wrong. The clutch pack ring thickness should have been checked prior to this step.
Answer B is wrong. The clutch pack reaction plate thickness also should have been performed prior to this step.
Answer C is wrong. Steel clutch plate thickness should have been measured before assembling the clutch drum.
Answer D is correct.

Answers to the Test Questions for the Additional Test Questions Section 6

1. C
2. A
3. C
4. C
5. C
6. D
7. B
8. D
9. B
10. A
11. C
12. A
13. C
14. A
15. D
16. D
17. C
18. A
19. A
20. C
21. B
22. B
23. B
24. D
25. C
26. A
27. D
28. B
29. C
30. A
31. C
32. D
33. A
34. C
35. B
36. A
37. B
38. C
39. C
40. A
41. D
42. B
43. C
44. C
45. C
46. C
47. A
48. C
49. A
50. D
51. D
52. A
53. D
54. C
55. A
56. A
57. A
58. A
59. C
60. A
61. A
62. A
63. C
64. B
65. A
66. D
67. D
68. C
69. B
70. B
71. C
72. D
73. B
74. A
75. B
76. B
77. B
78. B
79. A
80. B
81. B
82. A
83. C
84. B
85. B
86. B
87. C
88. A
89. D
90. D
91. B
92. B
93. B

Explanations to the Answers for the Additional Test Questions Section 6

Question #1
Answer A is wrong.
Answer B is wrong.
Answer C is correct.
Answer D is wrong.

Question #2
Answer A is correct.
Answer B is wrong. A dial indicator is used to measure end-play.
Answer C is wrong.
Answer D is wrong.

Question #3
Answer A is wrong.
Answer B is wrong.
Answer C is correct.
Answer D is wrong.

Question #4
Answer A is wrong.
Answer B is wrong.
Answer C is correct.
Answer D is wrong.

Question #5
Answer A is wrong. Low battery voltage will affect shift quality.
Answer B is wrong. A defective vehicle speed sensor will affect shift quality.
Answer C is correct.
Answer D is wrong. A poor ground will affect shift quality.

Question #6
Answer A is wrong.
Answer B is wrong.
Answer C is wrong.
Answer D is correct.

Question #7
Answer A is wrong. The bolts may be stripped.
Answer B is correct.
Answer C is wrong. Loose bolts often cause flexplates to crack.
Answer D is wrong. Loose bolts often cause convert pilot damage.

Question #8
Answer A is wrong.
Answer B is wrong.
Answer C is wrong.
Answer D is correct.

Question #9
Answer A is wrong. The engine draws vacuum, therefore oil would have to be drawn into the engine from the vacuum modulator.
Answer B is correct.
Answer C is wrong.
Answer D is wrong.

Question #10
Answer A is correct.
Answer B is wrong. The parking pawl gear engages the outer diameter of the ring gear.
Answer C is wrong. The parking pawl gear engages the outer diameter of the ring gear.
Answer D is wrong. The parking pawl gear engages the outer diameter of the ring gear.

Question #11
Answer A is wrong.
Answer B is wrong.
Answer C is correct.
Answer D is wrong.

Question #12
Answer A is correct.
Answer B is wrong. Seals should never be installed dry; installing seals dry can cause them to rip or tear during installation.
Answer C is wrong.
Answer D is wrong.

Question #13
Answer A is wrong.
Answer B is wrong. B
Answer C is correct.
Answer D is wrong.

Question #14
Answer A is correct.
Answer B is wrong. Chain stretch is not measured with a dial indicator.
Answer C is wrong.
Answer D is wrong.

Question #15
Answer A is wrong. Compression of the clutch piston return spring is not heard during air checks.
Answer B is wrong. Clutch packs are air pressure tested.
Answer C is wrong. If the check ball is missing no solid clunk will be heard.
Answer D is correct.

Question #16
Answer A is wrong.
Answer B is wrong.
Answer C is wrong.
Answer D is correct.

Question #17
Answer A is wrong.
Answer B is wrong.
Answer C is correct.
Answer D is wrong.

Question #18
Answer A is correct.
Answer B is wrong. Input shaft bearing failure is less likely than case wear.
Answer C is wrong. Upshots are less likely to be affected.
Answer D is wrong. Driveling vibration is not likely from improper end play.

Question #19
Answer A is correct.
Answer B is wrong. Transmission lines should be spliced with double flare unions.
Answer C is wrong.
Answer D is wrong.

Question #20
Answer A is wrong.
Answer B is wrong.
Answer C is correct.
Answer D is wrong.

Question #21
Answer A is wrong. Constant velocity joints generally make noise on a turn; thus the noise described in this question is probably coming from something else.
Answer B is correct.
Answer C is wrong.
Answer D is wrong.

Question #22
Answer A is wrong. A battery should have 9.6 volts at the completion of a battery capacity test.
Answer B is correct.
Answer C is wrong.
Answer D is wrong.

Question #23
Answer A is wrong. The gasket should not cover separator plate holes.
Answer B is correct.
Answer C is wrong.
Answer D is wrong.

Question #24
Answer A is wrong. A defective transmission case is not the most likely cause for the cracks at the bell housing bolt holes.
Answer B is wrong. Use of a pry bar may be required to separate the transmission and engine block once the bell housing bolts are removed.
Answer C is wrong. The flexplate and converter are marked to help maintain correct balance.
Answer D is correct.

Question #25
Answer A is wrong. The case cannot be honed.
Answer B is wrong. The case cannot be honed.
Answer C is correct.
Answer D is wrong. Metal rings can not be used as a substitute.

Question #26
Answer A is correct.
Answer B is wrong. The Technician is measuring pump gear-to-pocket clearance.
Answer C is wrong. The Technician is measuring pump gear-to-pocket clearance.
Answer D is wrong. The Technician is measuring pump gear-to-pocket clearance.

Question #27
Answer A is wrong. Not all transmissions have adjustable bands.
Answer B is wrong. Band adjustment bolts must be tightened then backed off a certain amount as specified in the service manual.
Answer C is wrong.
Answer D is correct.

Question #28
Answer A is wrong. Throttle linkage does modify throttle pressure.
Answer B is correct.
Answer C is wrong. Throttle linkage does modify shift points.
Answer D is wrong. Throttle linkage does indicate engine load.

Question #29
Answer A is wrong.
Answer B is wrong.
Answer C is correct.
Answer D is wrong.

Question #30
Answer A is correct.
Answer B is wrong. This component is likely to be damaged by metal contamination.
Answer C is wrong. This component is likely to be damaged by the broken tooth.
Answer D is wrong. This component is likely to be damaged by the broken tooth.

Question #31
Answer A is wrong.
Answer B is wrong.
Answer C is correct.
Answer D is wrong.

Question #32
Answer A is wrong. Clutch discs do need to be soaked in proper type fluid prior to assembly.
Answer B is wrong. All parts should be cleaned before assembly.
Answer C is wrong.
Answer D is correct.

Question #33
Answer A is correct.
Answer B is wrong. Governor pressure is not adjustable.
Answer C is wrong.
Answer D is wrong.

Question #34
Answer A is wrong.
Answer B is wrong.
Answer C is correct.
Answer D is wrong.

Question #35
Answer A is wrong. If you can squeeze fluid from a band it indicates there it still has some life left.
Answer B is correct.
Answer C is wrong.
Answer D is wrong.

Question #36
Answer A is correct.
Answer B is wrong. Clutch pack clearance is changed by installing varying thickness snap rings.
Answer C is wrong.
Answer D is wrong.

Question #37
Answer A is wrong. The throttle position sensor has no influence on unlocking the converter when stopping.
Answer B is correct.
Answer C is wrong.
Answer D is wrong.

Question #38
Answer A is wrong.
Answer B is wrong.
Answer C is correct.
Answer D is wrong.

Question #39
Answer A is wrong.
Answer B is wrong.
Answer C is correct.
Answer D is wrong.

Question #40
Answer A is correct.
Answer B is wrong. Governors decide at what point to shift the transmission.
Answer C is wrong. The governor does not control the firmness of the shifts.
Answer D is wrong. The torque converter is the only component that controls stall speed.

Question #41
Answer A is wrong. A damaged brake on/off switch may prevent torque converter clutch application.
Answer B is wrong. A damaged engine coolant temperature sensor can prevent torque converter clutch application.
Answer C is wrong. A damaged throttle position sensor can prevent torque converter clutch application.
Answer D is correct.

Question #42
Answer A is wrong. Discs in this condition should be replaced.
Answer B is correct.
Answer C is wrong.
Answer D is wrong.

Question #43
Answer A is wrong. This could be the cause.
Answer B is wrong. This could be the cause.
Answer C is correct.
Answer D is wrong. This could be the cause.

Question #44
Answer A is wrong.
Answer B is wrong.
Answer C is correct.
Answer D is wrong.

Question #45
Answer A is wrong.
Answer B is wrong.
Answer C is correct.
Answer D is wrong.

Question #46
Answer A is wrong.
Answer B is wrong.
Answer C is correct.
Answer D is wrong.

Question #47
Answer A is correct.
Answer B is wrong. The 2nd clutch is not applied in manual 1st.
Answer C is wrong.
Answer D is wrong.

Question #48
Answer A is wrong. The Technician is removing an extension housing seal.
Answer B is wrong. The Technician is removing an extension housing seal.
Answer C is correct.
Answer D is wrong. The Technician is removing an extension housing seal.

Question #49
Answer A is correct.
Answer B is wrong. The Technician is removing or installing a governor.
Answer C is wrong. The Technician is removing or installing a governor.
Answer D is wrong. The Technician is removing or installing a governor.

Question #50
Answer A is wrong. Transmission filters should not be reused.
Answer B is wrong. Some blackout deposits are the result of normal operation.
Answer C is wrong.
Answer D is correct.

Question #51
Answer A is wrong. Vane ends should be rounded.
Answer B is wrong. The vanes need replacement but the vane sizing is critical.
Answer C is wrong.
Answer D is correct.

Question #52
Answer A is correct.
Answer B is wrong. No hydraulic circuit is required to activate a one-way clutch.
Answer C is wrong.
Answer D is wrong.

Question #53
Answer A is wrong. Bands and servos do not have to be inspected if a one-way clutch has failed.
Answer B is wrong. A transmission cooling system failure would not cause a one-way clutch failure.
Answer C is wrong. The clutch pack is not part of the hydraulic circuit.
Answer D is correct.

Question #54
Answer A is wrong.
Answer B is wrong.
Answer C is correct.
Answer D is wrong.

Question #55
Answer A is correct.
Answer B is wrong. The low-reverse clutch is not applied in automatic 1st gear.
Answer C is wrong.
Answer D is wrong.

Question #56
Answer A is correct.
Answer B is wrong.
Answer C is wrong.
Answer D is wrong.

Question #57
Answer A is correct.
Answer B is wrong. Air pressure tests can check for correct servo seal installation.
Answer C is wrong. Air pressure tests can check for correct piston seal installation.
Answer D is wrong. Air pressure tests can check for correct hydraulic circuit blockages.

Question #58
Answer A is correct.
Answer B is wrong. Modulators are not adjusted using shims.
Answer C is wrong.
Answer D is wrong.

Question # 59
Answer A is wrong.
Answer B is wrong.
Answer C is correct.
Answer D is wrong.

Question # 60
Answer A is correct
Answer B is wrong. Some accumulator pistons can be installed upside down.
Answer C is wrong.
Answer D is wrong.

Question #61
Answer A is correct.
Answer B is wrong. A missing check ball in a shaft could cause damage to a one-way clutch or planetary gearset due to lack of lube pressure.
Answer C is wrong. An obstructed oil passage could cause damage to a one-way clutch or planetary gearset due to lack of lube pressure.
Answer D is wrong. A worn shaft bushing could cause damage to a one-way clutch or planetary gearset due to lack of lube pressure.

Question #62
Answer A is correct.
Answer B is wrong. Incorrect speed sensor readings will cause transmission shift concerns.
Answer C is wrong.
Answer D is wrong.

Question #63
Answer A is wrong. On a wavy snap ring you should measure from the top of the pressure plate to the bottom of the highest point of the snap ring wave.
Answer B is wrong. On a wavy snap ring you should measure from the top of the pressure plate to the bottom of the highest point of the snap ring wave.
Answer C is correct.
Answer D is wrong. On a wavy snap ring you should measure from the top of the pressure plate to the bottom of the highest point of the snap ring wave.

Question #64
Answer A is wrong. Bushings should not be heated with a torch.
Answer B is correct.
Answer C is wrong.
Answer D is wrong.

Question #65
Answer A is correct.
Answer B is wrong. Heat would distort the transmission case.
Answer C is wrong. A chisel would crack the transmission case.
Answer D is wrong. Snap ring pliers cannot be used to remove bushings; they are only used for snap rings.

Question #66
Answer A is wrong. Thrust washers often have selective thickness.
Answer B is wrong. Thrust washers often have oil passage holes.
Answer C is wrong. Thrust washers often have positioning tabs.
Answer D is correct.

Question #67
Answer A is wrong. The tools are set up to measure side gear backlash.
Answer B is wrong. The tools are set up to measure side gear backlash.
Answer C is wrong. The tools are set up to measure side gear backlash.
Answer D is correct.

Question #68
Answer A is wrong.
Answer B is wrong. B
Answer C is correct.
Answer D is wrong.

Question #69
Answer A is wrong. Bushings are usually a pressed in fit.
Answer B is correct.
Answer C is wrong.
Answer D is wrong.

Question #70
Answer A is wrong. Damaged transmission mounts on a rear wheel drive vehicle will not cause uneven tire wear.
Answer B is correct.
Answer C is wrong.
Answer D is wrong.

Question #71
Answer A is wrong.
Answer B is wrong.
Answer C is correct.
Answer D is wrong.

Question #72
Answer A is wrong. Higher throttle valve pressure results in shifts at higher vehicle speed.
Answer B is wrong. Shifts will occur at the specified speed when the throttle pressure is set correctly.
Answer C is wrong. Shifts would not occur at the same vehicle speed with misadjusted throttle valve pressure.
Answer D is correct.

Question #73
Answer A is wrong. A worn gearset would cause excessive noise, not burn the fluid.
Answer B is correct.
Answer C is wrong.
Answer D is wrong.

Question #74
Answer A is correct.
Answer B is wrong. The stall test should never take more than three seconds; power braking the vehicle for more than three seconds will overheat the transmission.
Answer C is wrong. The stall test should never take more than three seconds; power braking the vehicle for more than three seconds will overheat the transmission.
Answer D is wrong. The stall test should never take more than three seconds; power braking the vehicle for more than three seconds will overheat the transmission.

Question #75
Answer A is wrong. Typically when friction modifiers are depleted, the transmission fluid darkens from heat.
Answer B is correct.
Answer C is wrong.
Answer D is wrong.

Question #76
Answer A is wrong. Shaft end play can be measured before rebuilding to determine the amount of wear in the transmission.
Answer B is correct.
Answer C is wrong.
Answer D is wrong.

Question #77
Answer A is wrong. The throttle lever is moved by a cable or linkage.
Answer B is correct.
Answer C is wrong.
Answer D is wrong.

Question #78
Answer A is wrong. Loose drive studs are reason to replace.
Answer B is correct.
Answer C is wrong.
Answer D is wrong.

Question #79
Answer A is correct.
Answer B is wrong. The DVOM is not connected correctly to measure the voltage drop on the battery cable.
Answer C is wrong. The DVOM is not connected correctly to measure the resistance of the battery cable.
Answer D is wrong. The DVOM is not connected correctly to measure the voltage drop of the starter motor windings.

Question #80
Answer A is wrong. The case does not have to be flushed with transmission fluid, but it does have to be cleaned using a transmission parts washer/steam cleaner.
Answer B is correct.
Answer C is wrong.
Answer D is wrong.

Question #81
Answer A is wrong. The first diagnostic test should be a visual inspection.
Answer B is correct.
Answer C is wrong.
Answer D is wrong.

Question #82
Answer A is correct.
Answer B is wrong. Delayed or missed shifts can be caused by a variety of problems, including electrical problems.
Answer C is wrong.
Answer D is wrong.

Question #83
Answer A is wrong.
Answer B is wrong.
Answer C is correct.
Answer D is wrong.

Question #84
Answer A is wrong. Drive chains that are worn beyond specifications are changed when rebuilding the transmission.
Answer B is correct.
Answer C is wrong.
Answer D is wrong.

Question #85
Answer A is wrong. When performing a pressure test, use only low pressures.
Answer B is correct.
Answer C is wrong.
Answer D is wrong.

Question #86
Answer A is wrong. A dished drum generally will cause a band to wear on the outer edges, not a worn band strut.
Answer B is correct.
Answer C is wrong.
Answer D is wrong.

Question #87
Answer A is wrong. Burnt fluid is usually brown.
Answer B is wrong. Engine oil mixed with transmission fluid will not appear milky.
Answer C is correct.
Answer D is wrong. Aerated fluid is usually foamy.

Question #88
Answer A is correct.
Answer B is wrong. An open in a shift solenoid would only cause shift problems.
Answer C is wrong. A short to power would cause slipping and soft shifts.
Answer D is wrong. A short to a shift solenoid would cause only shifting problems.

Question #89
Answer A is wrong. Some contamination and black deposits are normal.
Answer B is wrong. Filters should be replaced not cleaned.
Answer C is wrong.
Answer D is correct.

Question #90
Answer A is wrong. Replacing the valve will not correct a damaged bore.
Answer B is wrong. Valve bodies should not be honed.
Answer C is wrong. A stronger spring will not correct the cause of the problem.
Answer D is correct.

Question #91
Answer A is wrong. Servo pistons must not be loose on their guide pins.
Answer B is correct.
Answer C is wrong.
Answer D is wrong.

Question #92
Answer A is wrong. A worn oil pump would cause low oil pressure at all engine speeds.
Answer B is correct.
Answer C is wrong. A stuck pressure regulator would cause a consistently high or low pressure.
Answer D is wrong. A plugged modulator hose would cause a high pressure all the time.

Question #93
Answer A is wrong. A slipping torque converter clutch would have no effect on engine RPM during a stall test.
Answer B is correct.
Answer C is wrong.
Answer D is wrong.

Glossary

AC An abbreviation for (1) alternating current. (2) air conditioning.

Aerated The introduction of air bubbles into the transmission fluid causing the fluid to expand and become compressible.

Alternating current (AC) The type of electrical current produced in an alternator.

Anodized To coat with electrolysis a metal surface with a protective oxide.

Arcing A term that applies to the spark that occurs when electricity jumps a gap.

Battery A device that stores electrical energy in chemical form.

Bench test A test or series of tests done by the technician on the work bench rather than on the vehicle.

Brakes The system that slows or stops a vehicle.

Bushing A sleeve, usually bronze, inserted into a bore to serve as a bearing for a rotating member.

Case grounded A term used when a component is grounded electrically through its case and is fastened directly to a metal, grounded part of the vehicle.

Chocked A term used when the wheels are blocked with safety chocks to insure safe working conditions on a running vehicle.

Clutch pack A series of clutch disc and plates that act as a driving or driven unit.

Computer An electronic device capable of following instructions and perform operations without human intervention.

Constant velocity joint One or more universal joints specially designed so acceleration-deceleration effects cancel each other out.

Control circuit A circuit that has the actual on/off responsibility in a circuit. It could be on the positive side or the negative side of the load, switch, or relay.

Coolant A fluid, usually a mixture of water and anti-freeze, used in the cooling system.

Crankshaft A revolving part in the lower section of the engine to which the connecting rods are attached.

DC An abbreviation for direct current.

Direct current (DC) A form of electrical energy produced by a battery.

Dressed A term used for the process of slightly altering a surface. Dressing can make a surface slightly smoother, rougher, or it can just remove slight imperfections.

Electrical A type of power produced by alternating or direct current.

Electrical harness A group of wires usually terminating at one or more connectors. Used to connect various components and systems.

Electronic That branch of science dealing with the motion, behavior, and emission of currents of free electrons.

Electronic control computer A digital device that controls engine and transmission functions electrically.

Electronic controls A term used for the electronic modules and associated components used to control many of the systems on today's vehicles.

Electronic transmission An automatic transmission that is computer controlled.

Elongated The stretching of the parts so they are no longer as designed. A hole that is out of round; egg-shaped.

End play The distance a shaft will move longitudinally.

Energized A term used to signify a component or circuit is activated electrically.

Erratic A term used when the operation of a component or system is not as designed or is intermittent.

Extension housing A housing that encloses the termination or output of a component. Like a transmission output shaft and its supporting bearings.

Fail-safe A term used for a backup system designed to protect certain systems on the automobile.

Fail-safe mode A special operating mode used to help eliminate complete break down of a system and allow the vehicle to reach a safe location.

Fault code Digits generated by the diagnostic programmer as an aid in troubleshooting.

Final drive The differential gears that provide power to the drive wheels.

Flexplate A slightly flexible steel part transferring power from the engine crankshaft to the torque converter.

Fuse An enclosure, usually in glass or plastic, of a calibrated link designed to open if a predetermined current is exceeded.

Fusible link A calibrated link designed to open if a predetermined current is exceeded.

Fuse link A calibrated link designed to open if a predetermined current is exceeded.

Garter spring A special spring placed behind the lip of some seals to insure good contact with the rotating part involved.

Governor A device that reacts to vehicle speed and enables up shifts and down shifts.

Ground A return path for an electrical current.

Ground-side switched A method of controlling current in an electrical circuit on the negative side of the component involved. Commonly used with electronic systems.

Half shaft joint A term used for a universal joint in the half shaft.

Hang-up A term used when something malfunctions, usually temporarily.

Heat exchanger An apparatus that allows heat to be transferred form one medium to another using the principle that heat moves to an object with less heat.

Hot-side switched A term used for the method of turning electrical and electronic circuits on/off from the positive side of the circuit.

Idle speed Speed of the engine, in RPM, at curb idle, under no load.

Ignition switch The main electrical switch of the vehicle, includes off, start, and run positions. Usually also includes an accessory (acc) position.

Inputs A term used with various electronic systems to identify the different signals sending information to the electronic modules about operating conditions.

Inch pound A unit of measure of torque in the English system.

Intake manifold A metal component used to direct and duct the air fuel mixture into the cylinders.

Intermittent A condition that occurs at random rather then as designed.

Load The demand for power placed on an engine.

Lapped A process of lightly smoothing or dressing a surface to check for straightness, flatness, or accuracy.

Line pressure The base pressure established in a transmission/transaxle by the pump and pressure regulator valve.

Lock-up converter a hydraulic torque converter that has friction material and a disc that can lock mechanically when commanded by an electronic module to improve fuel economy.

Lock-up solenoid an electrical solenoid that can enable or prevent torque converter lock-up.

Lockup torque converter A fluid-driven clutching assembly designed to improve automatic transmission coupling efficiency.

Lubrication Oil, grease, or other fluid used to coat moving parts to reduce friction.

Modulator valve A device located on the transmission that reacts to changing engine vacuum and controls the transmission throttle pressure according to engine load.

Neutralized To center or relieve stress on the power train mounts to minimize noise and vibration.

Neutralizing The centering of shafts and other structures to eliminate stress and/or rotating problems.

Nipple A term often used to signify a place to attach a fluid line, vacuum line, or vent hose.

Normally-open (no) A term used to signify the usual operating status of electrical devices or fluid control devices. When the proper command is received the device switches to closed.

Ohmmeter An electrical device used to measure resistance.

Oil cooler A radiator- or tank-like device used to cool engine, transmission, power steering, or other fluids.

Parking pawl A mechanical device required on automatic transmissions and transaxles to physically lock the output shaft and hold the vehicle in place when the shift lever is placed in park.

Planetary gear A gear set providing the means for neutral, low, medium, high, and reverse operation.

Preload A specified pressure applied to a part or assembly during assembly or repairs so the part can operate as designed.

Pressure test Varied methods on various systems to determine if components can hold pressure, or if the proper pressures are being developed and controlled in different automotive systems.

Radiator A heat exchanger in the engine cooling system.

Relay An electro mechanical switch.

RPM An abbreviation for Revolutions Per Minute.

Scanner A common name for various tools that can access information from different electronic control modules controlling electronic systems.

Selective thrust washer A special thrust washer that is furnished in various thicknesses so clearances and preload can be adjusted and changed if necessary.

Short circuit The intentional or unintentional grounding of a current carrying electrical wire.

Slips A term often used in the transmission area indicating a condition of lost efficiency. Bands slip, clutches slip, and one-way clutches slip.

Solenoid An electro-mechanical device used to impart a push-pull motion.

Speedometer A dash mounted device used to indicate road speed.

Speed sensor A device on today's vehicles that is an input to many of the different electronic systems. It usually relays component speed to the proper module with a sine wave type signal.

Stall test A test performed on automatic transmissions to determine engine and transmission condition.

Throttle valve A device connected the throttle linkage or the modulator valve and responds to engine load. The higher the demand the higher the pressure rises. It is often adjustable.

Torque The measure of a force producing tension and rotation around an axis.

Torque converter A unit that transmits power from the engine to the transmission.

Torque to yield To tighten to a specified predetermined yield or stretch point.

Torque wrench A tool used to measure and tighten a device to a specific torque.

Transmission fluid A lubricant formulated for use in transmissions.

Transmission modulator valve A device that regulates hydraulic pressure in the transmission.

Transmission oil A fluid formulated and designated for use in a transmission.

Transmission oil pan A removable part of the transmission that contains its oil.

TV An abbreviation for throttle valve. The throttle valve is connected the throttle linkage or the modulator valve and responds to engine load. The higher the demand the higher the pressure rises. It is often adjustable.

Warning lamp A light placed in the passenger compartment to warn the driver if a problem in a monitored circuit occurs.